Lab Manual to Accompany
Turfgrass Science and Management

THIRD EDITION

Join us on the web at

www.Agriscience.Delmar.com

Lab Manual to Accompany
Turfgrass Science and Management

THIRD EDITION

Robert D. Emmons

Developed by
Robert W. Boufford

Africa • Australia • Canada • Denmark • Japan • Mexico • New Zealand • Philippines
Puerto Rico • Singapore • Spain • United Kingdom • United States

NOTICE TO THE READER

Publisher does not warrant or guarantee any of the products described herein or perform any independent analysis in connection with any of the product information contained herein. Publisher does not assume, and expressly disclaims, any obligation to obtain and include information other than that provided to it by the manufacturer.

The reader is expressly warned to consider and adopt all safety precautions that might be indicated by the activities herein and to avoid all potential hazards. By following the instructions contained herein, the reader willingly assumes all risks in connection with such instructions.

The Publisher makes no representation or warranties of any kind, including but not limited to, the warranties of fitness for particular purpose or merchantability, nor are any such representations implied with respect to the material set forth herein, and the publisher takes no responsibility with respect to such material. The publisher shall not be liable for any special, consequential, or exemplary damages resulting, in whole or part, from the readers' use of, or reliance upon, this material.

Delmar Staff
Business Unit Director: Susan L. Simpfenderfer
Executive Editor: Marlene McHugh Pratt
Acquisitions Editor: Zina M. Lawrence
Developmental Editor: Andrea Edwards Myers
Executive Marketing Manager: Donna J. Lewis
Executive Production Manager: Wendy A. Troeger
Production Editor: Carolyn Miller

COPYRIGHT © 2000 Delmar, a division of Thomson Learning, Inc. The Thomson Learning™ is a trademark used herein under license.

Printed in the United States of America
4 5 6 7 8 9 10 XXX 06 05 04 03

For more information, contact Delmar, 3 Columbia Circle, PO Box 15015, Albany, NY 12212-0515; or find us on the World Wide Web at http://www.delmar.com

ALL RIGHTS RESERVED. No part of this work covered by the copyright hereon may be reproduced or used in any form or by any means—graphic, electronic, or mechanical, including photocopying, recording, taping, Web distribution or information storage and retrieval systems—without the written permission of the publisher.

For permission to use material from this text or product contact us by Tel (800) 730-2214; Fax (800) 730-2215; www.thomsonrights.com

ISBN: 07668-1553-6

Library of Congress Catalog Card Number: 99-051446

Contents

List of Figures viii

List of Tables x

Preface xi

Unit 1: Introduction 1
    TURFGRASS MANAGEMENT 1
    THE TURFGRASS MANAGEMENT COURSE 1
    USING THIS MANUAL 2
    THE COMPUTER: A TURF MANAGEMENT TOOL 2
    ACTIVITIES 4
    EXERCISES 5

Unit 2: Turf Information Management 7
    FINDING AND ORGANIZING TURF INFORMATION 7
    INFORMATION FOR MANAGING A TURF AREA 10
    ACTIVITIES 11
    EXERCISES 14

Unit 3: Turfgrasses 17
    TURFGRASS PLANT ANATOMY 17
    TURFGRASS IDENTIFICATION 19
    COMMON TURFGRASSES 21
    ACTIVITIES 23
    EXERCISES 30

Unit 4: Turf Soils 33
    TURF SOIL PHYSICAL PROPERTIES 33
    TURF SOIL CHEMISTRY AND FERTILITY 36
    SOIL TESTING AND RECORDS 37
    ACTIVITIES 39
    EXERCISES 44

Unit 5: Turf Establishment 45
    SELECTING A TURFGRASS 45
    SITE PREPARATION 48
    ESTABLISHMENT BY SEED 49
    VEGETATIVE ESTABLISHMENT 50
    CALCULATING MATERIALS NEEDS 51
    POST-PLANTING CARE 53
    ACTIVITIES 54
    EXERCISES 55

Unit 6: Fertilization **58**
- DETERMINING FERTILIZER NEED 58
- FERTILIZER MATERIALS 62
- SELECTING FERTILIZER MATERIALS 63
- TURF FERTILIZER APPLICATION 64
- ACTIVITIES 66
- EXERCISES 68

Unit 7: Mowing **70**
- MOWING PROGRAM 70
- MOWING EQUIPMENT 73
- MOWING COSTS 75
- ACTIVITIES 78
- EXERCISES 81

Unit 8: Irrigation **84**
- DETERMINING THE NEED TO IRRIGATE 84
- IRRIGATION SCHEDULING 87
- UNIFORM WATER APPLICATION 90
- ACTIVITIES 91
- EXERCISES 94

Unit 9: Turf Pest Problems **96**
- TURF WEEDS 96
- TURF INSECTS AND RELATED PESTS 97
- TURF DISEASES 100
- ACTIVITIES 102
- EXERCISES 103

Unit 10: Turf Pest Control **104**
- PEST CONTROL THROUGH PROPER CULTURE 104
- MONITORING FOR PEST PROBLEMS 104
- NONCHEMICAL PEST CONTROL 105
- BIOLOGICAL PEST CONTROL 105
- CHEMICAL PEST CONTROL 106
- ACTIVITIES 108
- EXERCISES 108

Unit 11: Turf Materials Application **111**
- FERTILIZER AND PESTICIDE APPLICATION METHODS 111
- CALIBRATION 113
- SPREADER CALIBRATION 113
- SPRAYER CALIBRATION 116
- ACTIVITIES 117
- EXERCISES 118

Unit 12: Turf Cultivation and Renovation ... **119**
    CULTIVATION ... 119
    TOPDRESSING ... 121
    THATCH CONTROL AND REMOVAL ... 121
    RENOVATION ... 123
    ACTIVITIES ... 123
    EXERCISES ... 125

Unit 13: Turf Equipment ... **126**
    EQUIPMENT SERVICE AND MAINTENANCE ... 126
    EQUIPMENT RECORD KEEPING ... 128
    REPLACING EQUIPMENT ... 128
    ACTIVITIES ... 129
    EXERCISES ... 130

Unit 14: Turf Management Practices ... **132**
    MANAGEMENT PLAN ... 132
    TURF RECORD KEEPING ... 134
    TURF MAINTENANCE SCHEDULE ... 134
    TURF BUDGET ... 134
    TURF MANAGEMENT SYSTEMS ... 135
    ACTIVITIES ... 136
    EXERCISES ... 136

Unit 15: Turf Communications and Publications ... **137**
    ORAL COMMUNICATION ... 137
    WRITTEN COMMUNICATION ... 138
    EXERCISES ... 139

Appendix A: Introduction to Computers ... **141**

Appendix B: Using the On-Line Companion ... **147**

Glossary ... **150**

Reference Library ... **155**

List of Figures

Figure 1-1	Large computers of a few years ago now fit on a small chip	3
Figure 1-2	Computer software inventory table	5
Figure 1-3	Computer hardware inventory table	6
Figure 2-1	Sample index card for organizing bibliographical information	10
Figure 2-2	Library search results	12
Figure 2-3	Sample turf reference database record	15
Figure 2-4	Sample site map	15
Figure 2-5	Table of area measurements from sample site map	16
Figure 3-1	Grass seed structures	17
Figure 3-2	Major vegetative grass plant structures	18
Figure 3-3	Grass leaf structures	20
Figure 3-4	Turfgrass adaptation zones of the United States	21
Figure 3-5	Seed structures table	23
Figure 3-6	Vegetative plant structures table	24
Figure 3-7	Cool-season grass seed samples table	26
Figure 3-8	Warm-season grass seed samples table	26
Figure 3-9	Seed mixtures table	27
Figure 3-10	Turfgrass identification table	27
Figure 3-11	Herbarium sheet (turfgrass plant collection card)	30
Figure 3-12	Sample turfgrass data form	31
Figure 3-13	Sample turfgrass cultivar record	32
Figure 4-1	Soil texture triangle	34
Figure 4-2	Availability of nutrients at various pH levels	38
Figure 4-3	Determining soil texture using a quart jar	39
Figure 4-4	Infiltration rates of tested areas table	40
Figure 4-5	Sample soil test submission form	42
Figure 4-6	Basic soil test data table	44
Figure 5-1	Establishment activities record	54
Figure 5-2	Seed application uniformity table	55
Figure 5-3	Sample turfgrass species/cultivar record	56
Figure 5-4	Sample establishment calculations form	57
Figure 6-1	Sample soil test results	60
Figure 6-2	Typical fertilizer label information	63

List of Figures

Figure 6-3	Fertilizer products table	67
Figure 6-4	Sample fertilizer calculations table	69
Figure 6-5	Sample fertilizer costs table	69
Figure 7-1	Example of one-third rule	72
Figure 7-2	Scissors-like cutting action of reel mowers	73
Figure 7-3	Chopping-like cutting action of rotary mowers	74
Figure 7-4	Mower operation and evaluation table	78
Figure 7-5	Mowing height evaluation table	78
Figure 7-6	Quality ratings (1 to 10) of rotary versus reel mower	79
Figure 7-7	Mower speed table	79
Figure 7-8	Estimated mowing time table	80
Figure 7-9	Actual turf mowing times table	81
Figure 7-10	Sample mower equipment record	82
Figure 7-11	Sample mowing time/cost calculation form	82
Figure 8-1	Tensionmeters are useful in measuring soil water level	86
Figure 8-2	Sample water budget worksheet	88
Figure 8-3	Water budget in checkbook ledger format	88
Figure 8-4	Deep rooting minimizes loss by deep percolation	90
Figure 8-5	Sample irrigation uniformity percentage calculation	91
Figure 8-6	Irrigation uniformity data tables	93
Figure 9-1	Comparison of grassy weeds and broadleaf weeds	97
Figure 9-2	Major structures of an insect	98
Figure 9-3	Example of complete metamorphosis	99
Figure 9-4	Example of simple metamorphosis	99
Figure 9-5	Typical chewing and piercing-sucking mouthparts	100
Figure 9-6	The disease triangle	101
Figure 10-1	Sample pesticide inventory form	108
Figure 10-2	Sample pesticide application record	109
Figure 11-1	Two major granular spreaders for turf use	112
Figure 11-2	Sample spreader calibration record	118
Figure 12-1	Two types of core aerifiers	120
Figure 12-2	A mechanical topdresser	121
Figure 12-3	Cross-section of a turf showing thatch layer	122
Figure 13-1	Comparison of mower maintenance on turf quality table	130

List of Tables

Table 2-1	Recommended Geometric Shapes with Area Formulas	14
Table 3-1	Common Cool-Season Turfgrasses	22
Table 3-2	Common Warm-Season Turfgrasses	23
Table 3-3	Vegetative Identification Key	28
Table 4-1	USDA Soil Separate Classification	33
Table 4-2	The Sixteen Elements Necessary for Plant Growth	36
Table 4-3	Calculating Soil Texture Percentages	40
Table 5-1	Characteristics of Common Cool-Season Turfgrasses	46
Table 5-2	Characteristics of Common Warm-Season Turfgrasses	47
Table 5-3	Seeding Rates of Major Turfgrass Species Including Seed Numbers	52
Table 5-4	Percent Species by Weight Versus Percent Species by Seeds per Pound	53
Table 6-1	Nutrient Deficiency Symptoms	59
Table 6-2	Recommended Nitrogen Rates for Most Turfgrasses	62
Table 7-1	Suggested Mowing Heights Based on Turf Use	71
Table 7-2	Suggested Mowing Height Ranges of Turfgrass Species	71
Table 9-1	Common Turfgrass Weeds	97
Table 9-2	Common Turf Insect Problems	98
Table 9-3	Common Turf Diseases	101

Preface

This manual is a companion guide to provide hands-on lab activities in a turfgrass management course. Along with being a traditional lab manual, it will hopefully serve as a reference after the student completes the course. A majority of the activities and exercises are typical tasks and practices performed almost daily by turf managers. A key feature of this manual is the exercises that a student can accomplish with traditional paper and pencil or with the use of computers.

Why computer-based exercises? Computers are becoming an integral part of many turf businesses. Business and industry leaders are recommending a knowledge of computers and computing as desired traits of the landscape or turf graduate. Many programs have addressed this by requiring a general computer literacy course.

Like English and similar communications courses, however, computer literacy courses, even those that use horticulture examples, are artificial. One of the objectives of this lab manual is to promote a "computing-across-the-curriculum" concept. This concept is similar to the "writing-across-the-curriculum" concept, which brings writing into technical courses so that students can apply writing to situations they will come across in their careers. The author hopes students will gain experience in using computers in turf management that they can carry with them into their careers.

This manual contains 15 units with each unit consisting of:

- a discussion to provide background information
- activities to be performed during the scheduled lab sessions
- exercises to be done by the students outside of regular classtime

The unit sequence is similar to the chapter sequence found in *Turfgrass Science and Management*, Third Edition by Robert Emmons. However, the units are general enough for use with any turf management textbook or as a stand alone resource guide.

Unit 1 includes a general introduction to turfgrass management, instructions for using the lab manual, and a rundown of additional materials the student will need for completing the various activities and exercises. The rest of the unit is a quick summary of computer use in turfgrass management. If students are not familiar with computers, they can refer to Appendix A for a short discussion of this topic. The other 14 units cover various topics related to turf science, culture, and management.

To assist students who elect to do the computer option of the various exercises, a supplemental on-line companion can be accessed through the web at www.Agriscience.Delmar.com. The files were developed using Microsoft® Works 3.0 for Windows. The author recommends that students have access to Microsoft® Works 3.0 for Windows. To accommodate the wide variety of software used in computer labs, an additional set of files in an "import" format is included. Please refer to Appendix B: Using the Supplemental On-Line Companion for further details.

ACKNOWLEDGMENTS

The author expresses appreciation to Robert Emmons and Delmar for the use of many of the illustrations in this manual. Appreciation also goes to Andrea Edwards Myers, Developmental Editor, Carolyn Miller, Production Editor, and the rest of the staff of Delmar for their assistance.

UNIT 1

Introduction

TURFGRASS MANAGEMENT

Turfgrasses compose the largest portion of vegetative ground covers in most landscape situations. Turfgrasses are also an integral part of the playing surfaces for many outdoor sports. For utilitarian areas such as roadsides and right-of-ways, turfs are the most effective groundcover for controlling erosion and providing safety zones.

A **turf** is composed of turfgrasses—grasses adapted for use in turfs. **Turfgrass science** is the study of these grasses and their use in a turf situation. **Turfgrass culture** or turfgrass maintenance is the science and practice of establishing and maintaining turfgrasses in a turf situation. **Turfgrass management** refers to turfgrass culture combined with the related business practices required to manage turfs in a facility. A turf manager is the person responsible for the care and management of turfs in a facility.

When a researcher looks at the impact of various mineral nutrient levels on turfgrass growth, the realm is that of turfgrass science. When a homeowner uses the information from turfgrass science research to determine the amount of fertilizer for application to a turf, the realm is that of turfgrass culture. Determining the cost of fertilizer as part of a budget, and assigning employees to apply the fertilizer compose the business activities related to turfgrass culture, that is, turfgrass management.

THE TURFGRASS MANAGEMENT COURSE

A successful turfgrass manager maintains a quality turf using the most effective cultural practices, and at the same time manages the turf in a way to keep expenses at a minimum. One of the goals of a course in turfgrass management is to provide the science behind the cultural practices that are necessary to maintain a quality turf. An understanding of the science behind the practices helps the turf manager effectively manage a turf area.

Turfgrass Management Labs

In any turfgrass management course, lab sessions are an important part of learning about turfgrass management. These sessions extend the topics discussed in lecture to actual turf situations. Because a turfgrass management course is an applied horticulture course, the labs can also provide hands-on experience in preparation for future job experiences.

A major goal of this lab manual is to provide activities that you, the student, can use in your career. After completing the turfgrass management course, you can continue to use many of the activities and exercises. The author hopes this lab manual will serve as a reference in your career.

Computers in Turf Management Labs

As you embark on your career, you will find that computers are becoming an integral part of many turf businesses. Business and industry leaders are recommending a knowledge of computers and computing as desired traits of the landscape or turf graduate. Many programs have addressed this by requiring a general computer literacy course.

Like English and similar communications courses, computer literacy courses, even those that use horticulture examples, are artificial. Your best writings on turfgrass management will be during and after this turf management course, when you are writing about topics related to turf management. Just as with writing, the best way for you to learn about computers is as they apply to turf management; you will do this in the turf management labs.

USING THIS MANUAL

This manual contains fifteen units: this **Introduction** and fourteen additional units, each covering a major topic of turfgrass management. The units are in a sequence very similar to many turfgrass management textbooks. Your instructor may change the unit sequence based on the course textbook and the time of year the course is being offered.

Discussions, Activities, and Exercises

Each unit consists of three parts: discussion, activities, and exercises. The discussion provides information in preparation for the activities and exercises. The discussion also helps review and enhance turf management lectures and readings.

You will perform the activities during lab sessions under the guidance of your instructor or lab assistant. Most of the activities involve practices and equipment used by turf managers in their daily responsibilities of managing turf areas. Since many of the lab activities will be outdoors, you should plan to dress for the weather.

You will do the exercises apart from scheduled lectures and lab sessions. The purpose of the exercises is to provide continued study and practice of the various turf-management activities learned in the lab sessions.

To provide learning situations with computers, some of the activities and a majority of the exercises have a computer-based option. Often, each exercise will consist of two options: a traditional paper-based option and an equivalent computer-based option.

The option you choose will depend on the availability of computer systems and on the specific assignments given by your instructor. Please visit www.Agriscience.Delmar.com for the supplemental on-line companion.

Additional Materials

In addition to this manual, you will need a three-ring binder, a calculator, and computer disks.

You will use the binder to hold all the paper materials created in the lab activities and exercises. It can also be used to hold your lab manual and other materials, to maintain an organized reference set.

The calculator will be used for various math calculations during laboratory activities in the field as several of the activities require immediate calculating to complete the activity. A basic calculator with arithmetic functions is more than sufficient for the activities. The calculator will also be useful in completing many of the paper-option exercises.

The computer diskettes are for storing files created in the various computer-based exercises. Even if you perform the exercises using a home computer, your instructor may require you to hand in completed assignments on diskettes.

THE COMPUTER: A TURF MANAGEMENT TOOL

A tool that is quickly becoming important to turf managers is the **computer**. The computer is important not only in gathering, organizing, and managing turf information; the computer also has become an important component in the cultural practices of maintaining a turf. For an increasing number of golf course superintendents and similar large-area turf managers, the computer is an integral part of the irrigation system. Other uses for the computer include environmental monitoring for potential pest problems and communicating with other turf managers.

A computer, in simple terms, is an electronic machine capable of following a list of instructions to perform calculations at a very rapid rate. What is more important, a computer is a multipurpose tool that enables the turf manager to store and process information in ways not easily done with paper and pencil.

A computer consists of hardware, or the actual machine, and software, or a list of instructions or programs that control the computer. The computer hardware and software, along with some related components, compose a computer system. A personal computer, or "PC," is small enough to fit on a desk; and many computers can not fit in the space of a notebook. For almost all turf management-related activities, the PC is the computer of choice.

Selecting a Computer System

When businesses first started using computers, the hardware was purchased first. This was because of the high cost of the equipment. Computer programmers then wrote software to accomplish various tasks. Each program was a handcrafted work of art to match the particular room-sized computer. Even including the labor required to write custom software, the cost of the hardware usually far exceeded any other costs.

With advantages in technology, computer hardware has become smaller (Figure 1-1), faster, and less costly to the point that the cost of software often exceeds the cost of hardware. This reversal has resulted in a change to implementing computer systems by way of a "solution-oriented" approach. The computer software is selected first, to meet an identified task. The requirements of the software then dictate the hardware purchased.

A similar situation has occurred in the golf turf industry. In the past, a tractor was the initial, major purchase (hardware), and various mower attachments (software) were subsequently bought to meet various mowing needs. Today, a golf course manager will first identify various mowing tasks and then purchase mowers specifically designed for those purposes, such as separate fairway mowers and rough mowers.

When selecting a computer system a turf manager should *DASSSH:*

*D*o an *A*nalysis of various turf activities
*S*elect *S*oftware to accomplish the various tasks done with computers
*S*elect *H*ardware required to run the selected software

FIGURE 1-1 Large computers of a few years ago now fit on a small chip. *(Courtesy of International Business Machines Corporation.)*

Selecting hardware first can be a costly mistake if the turf manager discovers there is little or no software to accomplish various turf management tasks.

ACTIVITIES

Introduction to the Turf Lab Facility

Many of the lab activities and exercises involve the use of existing turfs found on campus or in nearby areas. With your instructor, visit the areas to familiarize yourself with the facility. Meet with the manager and staff of the facility.

Record the following information for later reference:

- name of the facility
- address of the facility
- purpose of the facility
- turf manager's name
- turf manager's office phone number
- names of staff responsible for care of the turf

Using a Computer System

If you are unfamiliar with computers, read the Appendix A: Introduction to Computers before the laboratory session.

With the assistance of your instructor, learn the basic operation of the computer systems you will be using for later lab exercises. Learn the following procedures and activities to properly operate a computer.

- proper way to turn the computer on and off
- adjusting the monitor for viewing
- using the mouse (or other pointing device)
- inserting and removing floppy diskettes
- inserting and removing CD-ROMs
- starting and exiting operating systems
- logging onto and off of a network
- starting and exiting programs
- using Windows-type interfaces
  - working with icons
  - using pull-down menus
  - opening, closing, resizing, and moving windows
  - using scroll bars in windows
  - using dialog boxes with buttons, radio buttons, check boxes, fields, and other controls

Working with Major Software Applications

To gain experience in using the major software applications for later lab exercises, do the following:

- use a word processor to enter one page of lecture notes
- use a spreadsheet to total grades and calculate an average grade
- use a database manager to enter the names and addresses of five companies that serve the turf industry
- use a web browser to find three turf-related sites

Computer File Management

Learn the following procedures for managing computer files on hard disk drives and floppy diskettes:

- file naming requirements of the operating system
- creating new data files
- copying files between floppy diskettes and hard disk drives or network drives
    - copying one file
    - copying groups of files
- advanced backup and restoring of files

EXERCISES

Software Inventory

Fill in the computer software inventory table (Figure 1-2) detailing the various software programs that you will be using for later lab exercises. If most of the programs are part of an integrated software application, note each module that is available in the integrated application.

Word Processor:	
Spreadsheet:	
Database Manager:	
Telecommunications:	
Graphics—Drawing:	
Graphics—Painting:	
Graphics—Charting:	
Graphics—CAD:	
Games:	
Utilities:	
Web Browser:	

FIGURE 1-2 Computer software inventory table

System Inventory

With the assistance of your instructor, fill in the computer hardware inventory table (Figure 1-3). Detail the various components of the computer system that you will be using for later lab exercises.

Model:	
Manufacturer:	
CPU:	
Memory:	
Floppy Disk 1:	
Floppy Disk 2:	
Hard Disk 1:	
Hard Disk 2:	
CD-ROM:	
Pointer Device:	
Video Monitor:	
Network:	
Printer:	
Operating System:	

FIGURE 1-3 **Computer hardware inventory table**

UNIT 2

Turf Information Management

As a student, you should consider the turfgrass management course to be a foundation, or base of expertise, in activities related to the establishment and culture of turfgrasses and the management of turf areas. One difference between a mediocre turf manager and a good turf manager is the ability to gather new information for solving problems, managing the turf, and improving skills.

Turf managers must be able to manage a large amount of information. The key to an efficient turfgrass management operation is record keeping. Good record-keeping habits will lead to efficient turfgrass management with reduced costs. With current and future legislation concerning chemical applications, the turf manager must maintain detailed records on any chemical applied to a turf, including fertilizers.

FINDING AND ORGANIZING TURF INFORMATION

New information on turfgrass growth and management is being published daily. Trade journals, scientific journals, manufacturer literature, conference proceedings, extension bulletins, turf textbooks, and the popular press all contribute to the growing body of information related to turfgrass management. The Internet has opened up many information resources that at one time were available only to a few turf researchers. The turf manager must have the skills to filter this wealth of information and find only that information needed to manage the turf area.

Traditional information access methods include searching organized catalogs and indexes to find information related to the topic in question. New ways of searching for information include using computer technology to search catalogs and indexes on CD-ROMs and the Internet.

Turf Information Resources

The turf manager has access to numerous publications that provide extensive information on the growth, care, and management of turf. Many of these resources are often outside the realm of traditional indexing systems, and the turf manager must be diligent in order to find much of the information.

Trade Publications. Trade publications are magazines that provide articles of benefit to a specific trade or industry, in this case turfgrass management. Usually the articles provide up-to-date information on the latest turfgrass culture and management techniques. Many times, the articles will also cover nonturf issues, such as employee relations. These issues may not be important in the growing of grass, but they are very necessary to the long-term survival of a turf business. The publications are also the main method of advertising for many of the companies that provide service products to the turf industry. Qualified turf managers can receive many trade publications free of charge (the cost of the publication being supported by the advertisers).

Most trade publications focus on a particular segment of the turf industry, such as golf course management, grounds management, or lawn care. Much of the information presented in articles and advertisements is usable across many segments of the turf industry. Other segments of the turf industry will adopt many techniques and equipment first developed for use on the golf course and first published in a golf course-specific trade journal.

Scientific Journals. As turf scientists complete research on some aspect of turfgrass science and management, they publish their findings in scientific journals. To certify accuracy of their research findings, the manuscripts are peer reviewed by fellow scientists before publication.

The research information presented in scientific journals often serves as a foundation or precursor to new products or techniques for the management of turf. Many of the articles provide a better understanding of the growth of turfgrass, which can eventually lead to better management practices.

Extension Publications. The Extension Service located at a state's Land Grant University publishes numerous bulletins and similar publications that provide timely, up-to-date information to the turf manager. One such publication produced by many state turf Extension agents is a monthly bulletin of anticipated pest problems that are likely to occur within the next one or two months. This information allows the turf manager to appropriately plan pest control practices to coincide with the highest potential incidence of the pest.

Textbooks and Scholarly Books. There are several textbooks that provide comprehensive coverage on the science and management of turfgrasses used in a turf situation. Given that these books are written primarily to enhance learning in a turfgrass management course, turf managers can use many turf textbooks as references after they graduate from school.

Scholarly books, sometimes used as textbooks in advanced turfgrass management courses, provide in-depth information and discussion pertaining to turfgrass management. These books often will focus on a particular area of turf, such as turfgrass **diseases** or turfgrass ecology. These books enable turf managers to expand their knowledge of turfgrass management beyond the coverage provided in comprehensive turfgrass management textbooks.

Manufacturer's Publications. Even though biased toward their products, many manufacturers produce publications, such as newsletters, that provide information of benefit to the turf manager. For instance, the manufacturer will often provide product information in much greater degree than can be found on the package label. This allows the turf manager to better compare various products for use in managing the turf.

CD-ROM. A CD-ROM offers large amounts of storage on a single 4.5 inch (12 cm) disk. This allows the turf manager access to a large amount of information that takes up very little room in the office. Many encyclopedias, dictionaries, and other reference materials are now available on CD-ROM. Of particular interest to the turf manager is the publication of chemical and material safety data sheets (MSDS) on CD-ROM. Many regulations require the turf manager to have a MSDS on-site for every purchased chemical. Instead of trying to obtain these sheets from the different vendors, a turf manager can buy one CD-ROM containing the MSDS for every turf and ornamental chemical currently being manufactured. A subscription service allows for regular updates of new CD-ROMs, with the subscription service usually costing much less than obtaining the equivalent paper copies. The CD-ROM will often contain a searching system with a variety of searching strategies for rapid access to a particular chemical. This allows the turf manager to study a chemical for potential problems before buying the chemical.

On-Line Services. With a modem and telecommunications software, the turf manager can call to an on-line service to search a turf topic without ever leaving the office. Unlike the sequential searching through paper indexes, many on-line services will allow the turf manager to search across several indexes at one time.

Often, these on-line information services provide full text retrieval of articles. At the press of a key, the turf manager can transfer an article to the office computer or fax machine in a matter of minutes.

The Library. The library is an excellent starting point for finding information on a turfgrass management topic. The library is not only a collection of books and periodicals; it is a facility that provides several resources for accessing information on an infinite range of topics and interests. Cooperative ventures are being developed among libraries to extend these resources beyond the walls of the local library. Many libraries use interlibrary loan, facsimile machines, and the Internet as means of extending their services to the community. These cooperative ventures allow the turf manager to access a distant state agricultural university library from the local library, and, thus, obtain the benefits of both library systems.

Peer Resources. Other turf managers and related personnel create an informal but important information resource. Their experiences and expertise are valuable sources for answers to questions that a turf manager cannot find in any published resources. Unfortunately, time and distance often restrict face-to-face or even telephone contact. With a modem and appropriate networking software, the turf manager can connect to the Internet and communicate with other turf managers around the world.

Turf Reference System

The turf manager should maintain a reference file of newly published information. This allows for quick access to information when the need arises. A stack of old trade magazines could be considered part of the reference library. If no effort is made to catalog the information, however, time can be lost in trying to locate the information at a later time when the need arises.

Developing a Reference System. Building a reference system involves selecting materials that will best fit anticipated information needs. The particular job will dictate what materials will be important. A golf course superintendent will likely build a collection of materials related to managing turf found on golf courses. A lawn care applicator will likely include not only agronomic information related to lawn care, but also information on business management practices as they pertain to running a lawn care business. A college grounds manager's interest will include the care of hardscapes and sports fields.

Various textbooks can serve as starting points for a reference collection. Given that a book already contains an index, there is little need to further organize the information. And many basic turfgrass science principles and turfgrass management practices typically included in textbooks will remain up-to-date for many years.

Periodicals will compose the largest part of a turf manager's reference system. They provide up-to-date information on new turfgrass management practices and on a myriad of related issues, such as pesticide usage regulations and current turf industry trends.

Developing a Bibliography. As a reference system grows, so does the need to be able to quickly find information in the stacks of periodicals and books that will collect in a turf manager's office. To maintain some control over the growing amount of printed information, a turf manager should catalog and index a collection into a working bibliography. Even though bibliographies are typically associated with writing, a bibliography can also serve as a means of organizing information into a reference system that pertains to a turf manager's work.

Index cards are a traditional means of developing a bibliography. The author, title, publication details, and a brief description of the reference source are written on one index card (Figure 2-1). The cards can then be organized based on a group of topics or subjects.

```
Mowing

Budding, E. 1999. Effects of blade speed on mowing.
Mower Science. 88(6):45-56
Keys: mowing, reel, rotary

Article discusses the effects of various mower speeds
on quality of cut. Slower speeds tend to shread
turfgrass blades while higher speeds tend to bruise
turfgrass blades. For reel mowers, frequency of clip
is important in mower selection.
```

FIGURE 2-1 Sample index card for organizing bibliographical information

Using Computers. Many turf specialists now use computers as a means of developing and maintaining a reference file system. Bibliography or reference software allows one to easily search through a file in several different ways. Many systems allow for integration with CD-ROM or on-line indexes so that citations can be transferred into the reference system.

INFORMATION FOR MANAGING A TURF AREA

Site Inventory

The foundation for developing an efficient turfgrass management program is the site inventory. The turf manager needs to have an accurate, up-to-date record set of all aspects of a turf site. This includes area measurements, the specific turfgrasses in a turf site, soil analysis, irrigation system maps, underground utilities, hardscapes, and nonturf plantings.

All of this information is important in developing a management plan for the turf site. Much of this information is critical to the health and quality of a turf. As an example, lack of proper turfgrass identification along with inaccurate area measurements can lead an employee to miscalculating the amount of **herbicide** for application to a turf. This may result in application of excessive herbicide amounts on a susceptible turfgrass, which could lead to damage or death of the turf.

Mapping the Turf Site. A turf manager should maintain up-to-date drawings of all turf areas. Besides providing the information necessary to develop management plans, the drawings are useful in directing employees to locations where they need to work. It is much easier to repair a broken irrigation line if maps are available showing the location of the shut-off valve.

Most small sites, such as home lawns, will likely lack drawings of the site. Turf managers, such as lawn care technicians, will need to develop a detailed drawing of the site before developing a cultural plan for the home owner. Since the site is often small, the technician can easily accomplish detailing the site with a measuring tape and graph paper.

For large facilities with extensive turf areas, such as golf courses or corporate campuses, drawings may already exist from the original architectural plans and blueprints. Because they do not display "as-built" details, however, these plans are usually, to some degree, inaccurate. The turf manager must update the plans to reflect the current site. To obtain an accurate, as-built plan of the facility, the turf manager may need to use the service of a surveying company.

A turf manager can use computer-aided-design (CAD) or computer-aided-facilities-management (CAFM) applications for mapping a turf site. These tools serve as "visual information managers" for the storage and maintenance of maps, drawings, and blueprints, much as database manager software tracks numeric data and text information.

A major advantage of using a CAD or CAFM system is the ability to generate a number of different paper drawings from one set of electronic drawings. Also, from the one electronic drawing, a turf manager can produce several paper drawings at different scales to show an overall view of the turf area, as well as various detailed views.

Measuring Turf Areas. A key component of the site inventory is accurate measurement of all turf areas. Accurate measurements allow a turf manager to perform a large variety of management practices with greater efficiency, reduced waste, and at lower costs.

There are several different techniques for measuring turf areas. Many of these techniques require a minimal amount of equipment. Among the more advanced techniques is using CAD systems to calculate areas from existing drawings or aerial photographs. The services of a surveying company may be required to obtain the most accurate measurements along with detailed maps.

ACTIVITIES

Using the Library

With assistance from your instructor and the library staff, fill in Figure 2-2 with the following information:

1. Using your textbook as a guide, write the name of a turf-related topic about which you would like more information.
2. Using the *Library of Congress Subject Headings* books, list two subject headings you might check in the library's catalog system for information on your turf-related topic.
3. Using the school library's catalog system, do a search with the selected subject headings as search keys. List two books that might provide information on the topic. Include the "call number" for locating the books.
4. Using the *Encyclopedia of Associations* or similar reference, list the name and address of an association that might be able to provide information on your topic.
5. Using a periodical index, find two articles about your turf-related topic. For each article list the article title, periodical title, volume number, issue number, and page number(s). Also list the periodical issue used.

If your school library has network access, continue with the following:

1. Access an on-line catalog system outside of the school's catalog system. This can include a statewide on-line library catalog system or a catalog system at another school. Find and list two more books that might provide information on the topic. Include the call number for locating the books.
2. Access an on-line periodical index outside of any index system (CD-ROM, on-line) at your school and list two more articles about your topic. List the article title, periodical title, volume number, issue number, and page number(s).
3. If the school network has access to the World Wide Web on the Internet, use one of the many available web search engines to find two Web sites that may contain information relating to your topic. Write down the Uniform Resource Locator (URL) or Web address of each site.

Turf topic	
Book 1 title and call number	
Book 2 title and call number	
Association resource	
Turf-related article 1 from periodical index	
Turf-related article 2 from periodical index	
On-line catalog book 1 from off-campus site	
On-line catalog book 2 from off-campus site	
On-line periodical index article 1	
On-line periodical index article 2	
Turf-related World Wide Web address	

FIGURE 2-2 Library search results

Mapping Out a Turf Area

With assistance from your instructor, map out an assigned turf area using one or more of the following methods:

- measuring tape and graph paper
- general purpose drawing or CAD software

As an alternate activity, obtain existing blueprints or drawings of a selected turf site and do one or both of the following:

- using a digitizer tablet, trace over the paper drawing to create a CAD drawing of the site
- scan the drawing into the CAD program and trace over the scanned image to create a CAD drawing

Keep any paper drawings and printed CAD drawings in your three-ring binder. Store all CAD drawings on a floppy diskette.

Area Measurement Using Best Fit of Geometric Shapes

The surface area of many turfs can be determined by dividing the area into simple geometric figures. Once sectioned, the area of each geometric figure is calculated using the various area calculations for simple geometric shapes. This method can be applied in the field via the use of a measuring tape and wooden stakes. The recommended geometric shapes and associated area formulas are listed in Table 2-1.

TABLE 2-1 Recommended Geometric Shapes with Area Formulas

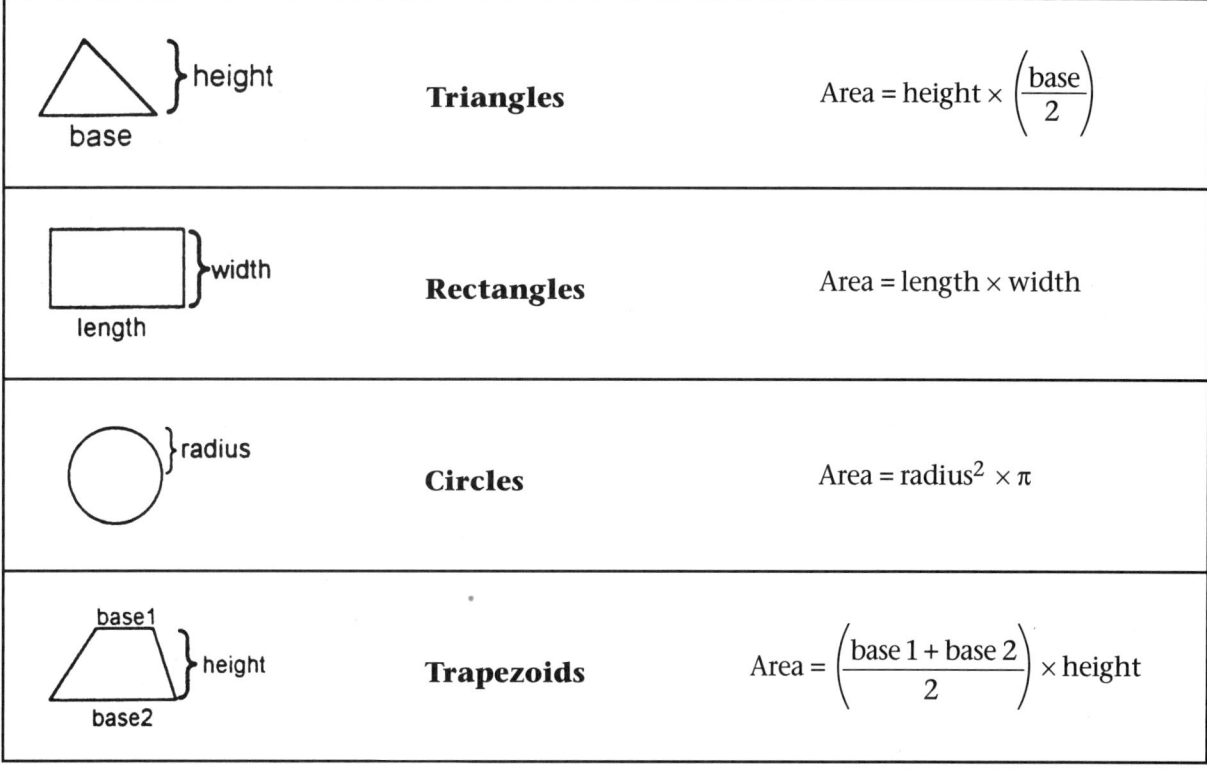

Using Paper. Determine the area of an assigned turf site using the best fit of geometric shapes. This can be done either in the field or from maps provided by your instructor. Record the area of each section on a data sheet for keeping in your three-ring binder. Include both square feet and acres (square meters and hectares). Be as accurate as possible because you will be using this information in later exercises.

Using a Computer. Obtain an accurate blueprint or drawing of the assigned turf area. Using a digitizer tablet and the appropriate area determination functions of a CAD program, trace the various turf areas to calculate the surface areas. As an additional activity, use the area of polygons functions of the CAD program to determine surface areas of an existing CAD drawing.

EXERCISES

Turf Reference File

Each week during the turfgrass management course, find three articles from magazines, trade journals, or scientific journals. These articles should relate to the topic that your instructor is discussing in the turfgrass management lecture. Include all information as shown in the sample index card (Figure 2-1).

Using Paper. Develop a paper record form or index card form with the following fields: Topic, Title, Journal, Year, Volume, Pages, Keywords, and Abstract. If you use ruled index cards, place the topic above the top red or double line. As you write out new records, organize the card file by topic.

Using a Computer. Develop a database record (Figure 2-3) using a database management program assigned by your instructor. Create the following fields: Topic, Title, Journal, Year, Volume, Pages, Keywords, and Abstract. If the database management program is capable of indexing, create an index on the Topic field.

On-line Companion. U02E01.WDB—Turf article reference database

Site Inventory Maps and Drawings

Map out a turf area assigned by your instructor. This area could be on your campus, on a golf course, on a municipal park, or a series of home lawns. Include all trees and shrub plantings along with all other landscape features. Be as detailed as possible because you will be using the information in later activities and exercises.

Using Paper. Develop maps of the site using grid paper. Store these drawings in your three-ring binder for use in later activities and exercises.

Using a Computer. Develop maps of the site (Figure 2-4) using a drawing program or a CAD program. Store the files on a floppy diskette or network for use in later activities and exercises. As a backup, print out the drawings for keeping in your three-ring binder.

On-line Companion. U02E02.WMF—Sample site drawing

Site Inventory of Area Measurements

Measure and calculate the areas of a selected turf site using the methods covered in the Activities section. This site could be your campus, a golf course, a series of home lawns, or a municipal park. Be as accurate as possible because you will be using the information in later activities and exercises.

Turf Information Management   15

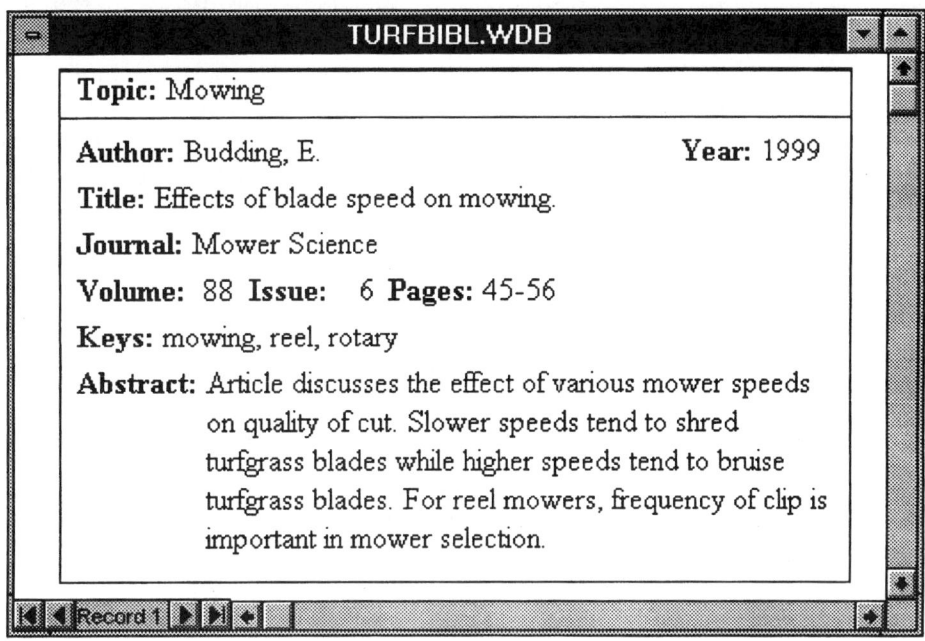

FIGURE 2-3 Sample turf database record

FIGURE 2-4 Sample site map

Using Paper. Develop a paper record in a table format. (The column-formatted sheets used by accountants will provide a table format for keeping the information.) Label all measured areas as reference keys to the maps. Keep the area list in your three-ring binder for use in later actitivites.

Using a Computer. Using either a spreadsheet program or a database program develop a table of area measurements (Figure 2-5). A spreadsheet will allow for later analysis, such as totaling all areas. The spreadsheet will permit easy extending of later calculations.

On-line Companion. U02E03A.WKS—Site inventory spreadsheet
U02E03B.WDB—Site inventory database

Hole Number	Area Description	Area (sq. ft.)
1	Tee	4,726
1	Green	9,248
1	Fairway	101,834
2	Tee	5,675
2	Green	7,179
2	Fairway	102,399
3	Tee	4,350
3	Green	6,300
3	Fairway	55,597
4	Tee	5,778
4	Green	7,235
4	Fairway	109,228
5	Tee	4,468
5	Green	6,399
5	Fairway	109,906
6	Tee	4,974
6	Green	8,841
6	Fairway	108,539
7	Tee	6,589
7	Green	6,585
7	Fairway	171,662
8	Tee	5,044
8	Green	8,579
8	Fairway	34,039
9	Tee	4,911
9	Green	9,882
9	Fairway	153,806
Total		1,063,773

FIGURE 2-5 Table of area measurements from sample site map

UNIT 3

Turfgrasses

TURFGRASS PLANT ANATOMY

A turf consists of a large population of individual turfgrass plants. It is this large population of plants cut at a uniform height that gives the smooth, carpet-like appearance of a turf. The structures or anatomy of a grass plant provides the underlying reason for many turfgrass management practices. For example, it is the short, stubby **crown** with short, compressed internodes of the turfgrass plant, that allows for routine mowing with limited impact on the growth and survival of the plant.

A knowledge of the different structures also provides a means of identifying the different grasses found in a turf. Many identification references, such as vegetative ID entries, depend on the turf manager having a working knowledge of grass structures. Misidentification of seeds could result in a poor quality turf from the wrong seed. Poor identification of the grasses in a turf stand can lead to an improper maintenance program, for example, applying a cultural regime for bluegrass when the turf stand contains mostly creeping bentgrass.

Turfgrass Seed

Grass seed structures are relatively simple (Figure 3-1). The lemma and palea that enclose the seed are leftover floral bracts, or modified leaves that originally surrounded an individual grass flower. A small, stub-like appendage at the base of the seed in front of the palea is the rachilla. This is the original point of attachment for the flower to the **inflorescence**. Some species of seed contain hair-like extensions known as awns. These extensions are useful in seed identification.

Turf Vegetative Structures

In addition to providing a means of identification, the vegetative structures of mature grass plants (Figure 3-2) serve a functional purpose. When looking at a turf, the leaves are the most visible structures. Leaves grow from nodes in the crown at ground level and extend either vertically or, under some conditions, more prostrate, to the ground surface.

Any vertical **stems** are part of the crown at ground level. For maintained turf, a stem appears only when the grass plant goes into flowering. Many times, mowing removes the flower head and stem. Many grasses used for a turf have lateral-growing stems, which improve the overall density of a turf.

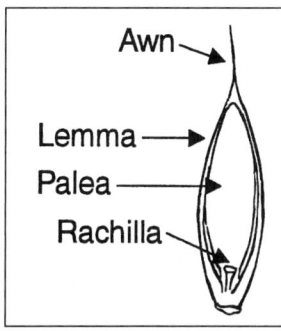

FIGURE 3-1 Grass seed structures

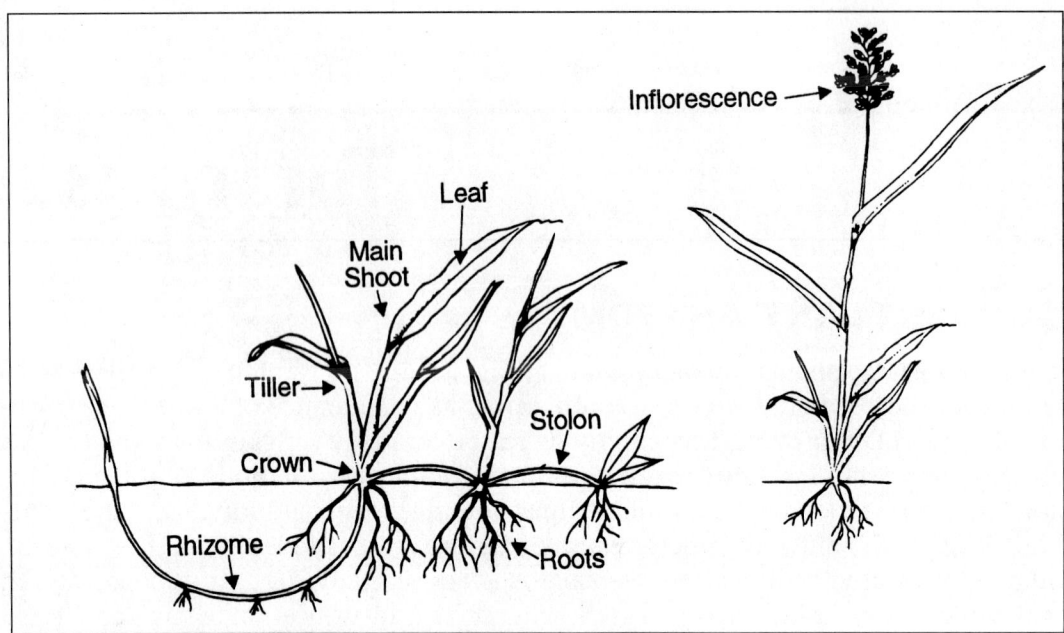

FIGURE 3-2 Major vegetative grass plant structures

Crown. The **crown** is the major structure of a turfgrass plant. The crown is a stem-type structure with short, compressed **internodes** and **nodes** containing leaves, axillary **buds**, and a growing point. The internodes in grasses used for turf usually do not elongate except during flowering. Adventitious roots emerge from the lower part of the crown; new shoots and lateral-growing stems emerge from the axillary buds in the nodes of the crown.

Tillers. **Tillers** are vertical-growing side shoots from axillary buds located at the nodes of the crown. Due to apical dominance, which is the influence of the main growing point of the crown over the growth of other buds and shoots, tillers are often smaller than the main shoot. Turfgrasses that produce only tillers exhibit a **bunch-type growth** habit.

Stolons. **Stolons** are lateral-growing stems that emerge from buds located at the nodes of the crown. These stems pierce the surrounding leaf **sheaths** and grow along the surface. New shoots and roots develop at nodes of these lateral-growing stems.

As with tillers, shoots that develop at the nodes of lateral stems are often smaller than the main shoot. To enhance growth and density, a turf manager will sever the internodes with vertical cutting machines. This severing removes the effects of apical dominance, resulting in better growth.

Rhizomes. **Rhizomes** are lateral-growing stems that emerge from buds located at the nodes of the crown. Unlike stolons, rhizomes will grow underground for a distance before emerging above the surface. Once the rhizome emerges above the surface, a new shoot forms from a terminal bud on the rhizome.

A turf with a rhizome growth habit can easily refill areas when chunks of turf, called divots, are torn from the turf. New growth will emerge from rhizomes crisscrossing just below the soil surface.

Many turfgrasses have either stolons, rhizomes, or both. Turfgrasses with these types of stem growth exhibit a **creeping-type growth habit.**

Inflorescence. The **inflorescence** is the flower of the grass plant. In a maintained turf, mowing will often remove the inforesence. With some grasses, the inflorescence stays below the mowing height. This tends to reduce the overall quality of a turf. Also, energies normally put toward shoot production are diverted to inflorescence growth, which may further reduce the quality of the turf.

TURFGRASS IDENTIFICATION

Proper identification of turfgrass seeds and plants is a very important part of turfgrass management. With proper identification of grasses in a turf, the turf manager can distinguish between desirable grasses and weedy grasses. The turf manager can then implement cultural practices to favor the desirable grasses while working to eradicate the undesirable grasses.

Turfgrass Seed Identification

Seed brought from a reputable supplier will usually come in sealed bags so that the purchaser can be certain of the species being bought. Open bags in a shop or open bins in a garden center, however, are subject to contamination from other seed mixes. Some of the seeds commonly used in turf are very similar in seed structure but very different in growth habit and use.

As an example, tall fescue (*Festuca arundinacea*) and fine fescue (*Festuca rubra*) are, on quick inspection, almost identical in seed size and characteristics. Tall fescue, however, has a wide, coarse leaf blade, while fine fescue has a narrow, almost bristle-like leaf blade. In many situations, tall fescue, when combined with other grasses, becomes a weed. This is because of its texture and its tendency to form unsightly clumps. Visual inspection and identification of turfgrass seed will satisfy the turf manager of an uncontaminated seed supply. This will safeguard against costly weed control practices in the future.

Identification of turfgrass seed involves study of the various characteristics of the seed. A key part of identification is noting the length, width, and shape of the seed. Other distinguishing characteristics include various features of the lemma and palea, along with the size and shape of the rachilla. The presence and length of awns can help in identifying the seeds of some grass **species**. Due to breakage during handling and shipping, the presence or absence of awns is not a reliable means of identification.

Turfgrass Plant Identification

Proper and continued identification of grasses in a turf environment is important in developing an appropriate management plan. Over time, the grass species in a turf population may change from the originally established grass species to a totally different grass species with different management requirements. Without proper and ongoing identification of grasses a turf manager will likely continue to apply the management program for the original grass species.

An example is the encroachment of annual bluegrass (*Poa annua*) into a creeping-bentgrass (*Agrostis palustris*) golf green. Too often, a superintendent will continue to apply a management program for creeping bentgrass even though annual bluegrass may have different watering, fertilization, and disease-control requirements. Continued evaluation of a turf area will assure both the turf manager and the customer that the turf area is comprised of quality turfgrass with no encroaching weedy grasses.

Along with the major structures of a grass plant, such as presence or absence of rhizomes and stolons, the structural parts of the leaf assist in identification of different turfgrass species. There are several distinguishing structures of a grass leaf that are very important in identifying grass species (Figure 3-3).

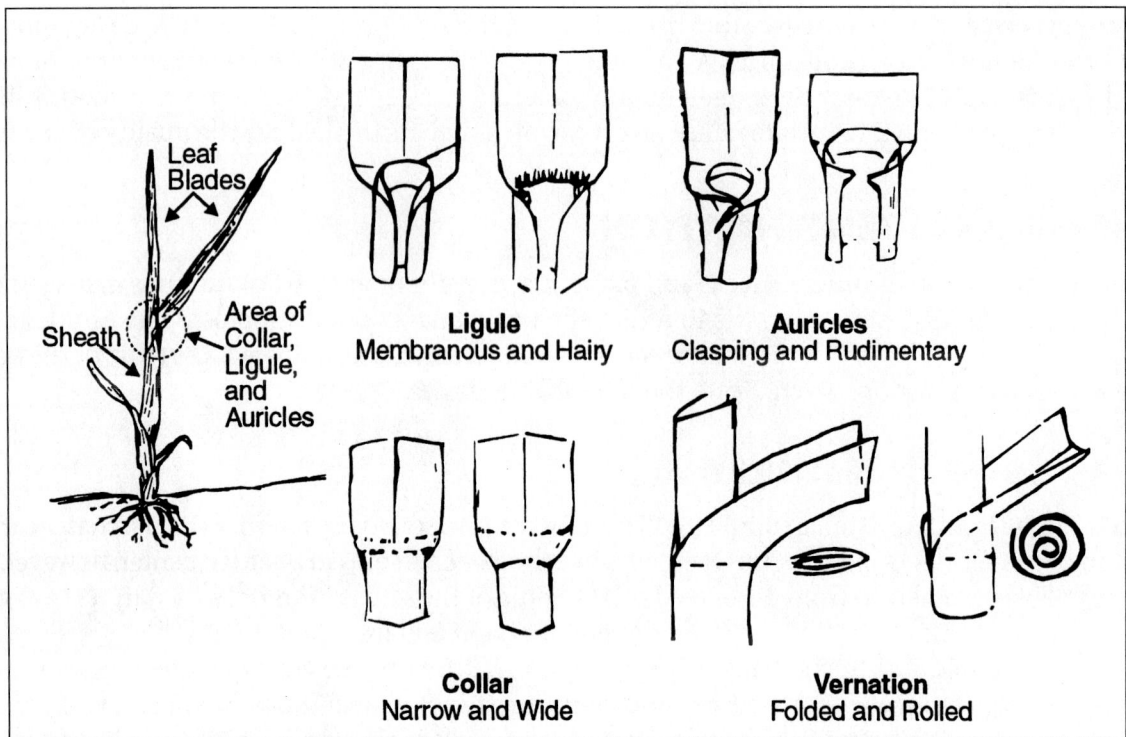

FIGURE 3-3 Grass leaf structures

Blade and Sheath. The entire grass leaf consists of the leaf blade and the leaf sheath. The leaf **blade** is the upper part of the leaf and is usually flat and open. The leaf blade is the most visible part of a grass plant in a maintained turf. The **sheath** is the lower part of the leaf, and is attached at a node at the crown. The sheath has a tube-like arrangement that provides support for the leaf blade.

Distinguishing characteristics of the leaf blade include the shape, width, tip shape, and numerous surface features. Among these surface features are pubescence (fine hairs), rough or smooth leaf blade edges, and prominent veins or ribs.

Like the blade, the sheath exhibits distinguishing characteristics such as hairs and the way in which its edges overlap to form a tube-like structure.

Collar. At the joint of the leaf blade and leaf sheath is the **collar**. Often, the collar is a lighter color than the surrounding leaf blade and sheath. One can best observe the collar from the back side of the leaf.

Different grasses exhibit varying collar widths and shapes. Some grass species have a fairly broad collar, while others exhibit a very narrow collar. Kentucky bluegrass (*Poa pratensis*) has a divided collar that appears to be in the shape of a bow tie.

Ligule. Just in front of the collar is the ligule. The **ligule** is just inside the throat formed by the sheath. The ligule is an eruption of epidermal tissue in front of the collar. The size and shape of the ligule are key features in distinguishing different grass species.

Ligules can be either **membranous** or a fringe of hairs. The ligula of centipedegrass (*Eremochloa ophiuroides*) is membranous with short hairs at the top.

Due to the small size of maintained turfgrass, a hand lens is useful in observing ligules. An easy means of observation is to hold the sheath between the thumb and forefinger of one hand. With the other hand, fold back the leaf blade at the collar until the blade touches the sheath. By holding the leaf blade and the sheath between the thumb and forefinger, the ligule will often stand erect for easier inspection.

Auricles. In some grasses, **auricles** are present at the collar. These finger-like extensions result from growth of the leaf margin of the collar. On some grasses, auricles are absent, while on others they are very prominent. The auricles can range from short and stubby (on tall fescue) to long and clasping (on annual ryegrass).

Vernation. The **vernation** is the arrangement of new leaves as they growth up through the leaf sheaths of older leaves. The youngest leaf will either be folded in half lengthwise (**folded vernation**) or rolled up (**rolled vernation**).

One must take care when observing vernation as a few common turfgrasses have a "folded-in-the-bud" vernation. After the leaf emerges, it has a tendency to roll up, giving the false impression of being a "rolled-in-the-bud" vernation.

COMMON TURFGRASSES

From a taxonomic standpoint, grasses belong to the Liliopsida (*Monocotyledoneae*), commonly referred to as **monocots**. Plants in the monocot class have leaves with parallel veins and a single cotyledon or "seed leaf." The grass family Poaceae (*Graminae*) consists of 600 genera with over 7,500 species. Of this large number of species, turf managers use only about 30 species in turfgrass culture. In a majority of turf situations, the number is much smaller, with fewer than 15 species used to a much greater degree than the others.

Within this small group of grasses, turfgrass species are further divided into groups based on optimum or ideal growing temperatures (Figure 3-4). The **cool-season** turfgrasses do best in temperatures between 60° and 75°F; **warm-season** turfgrasses do best in temperatures between 80° and 95°F.

Where grasses are used is determined by the average climate conditions, particularly the annual variation in temperature. Cool-season grasses will be found in the cooler regions of the country, commonly known as the *cool-season zone;* warm-season grasses will be found in the warmer climates of the *warm-season zone*.

Between the two zones is an arbitrary area approximately 50–200 miles wide known as the **transition zone**. Here, cool-season grasses grow best during the cooler winter months when warm-season grasses are dormant. In the summer heat, the reverse occurs; warm-season grasses grow at peak and cool-season grasses are slow growing or dormant. With intense care, particularly irrigation and pest control, many cool-season grasses can grow year-round in the lower-transition and warm-season

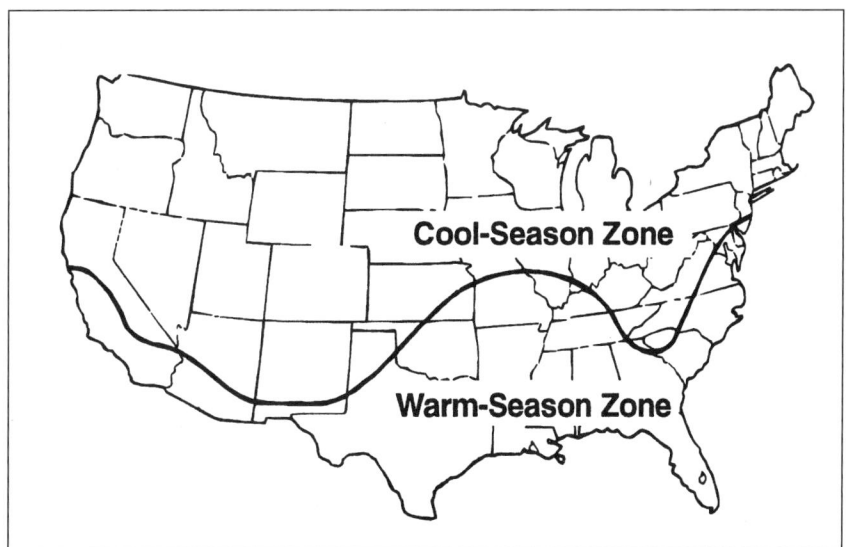

FIGURE 3-4 Turfgrass adaptation zones of the United States

zones. Because of their poor tolerance to below-freezing winter temperatures, however, warm-season grasses rarely survive the winters of the cool-season zone.

In addition to optimum growing temperatures, grasses are also classified into groups based on other growing characteristics, including growth habit and **leaf texture**. Growth habit, bunch-type or creeping-type, has an impact on a turf's ability to quickly repair itself when subject to tearing conditions, such as those that occur on a football field. Leaf texture has an impact on the visual quality of a turf. For the sports turf manager, density has an impact on the quality of the turf playing surface.

Cool-Season Turfgrasses

Of the cool-season grasses, Kentucky bluegrass (*Poa pratensis*) and perennial ryegrasses (*Lolium perenne*) are used in a wide variety of turf situations, including home lawns, commercial lawns, sports fields, and golf courses. These grasses are fairly compatible with regard to leaf texture and leaf color. Kentucky bluegrass has a rhizome growth habit, which enables the formation of a tight dense sod that is very stable under a variety of traffic conditions. Perennial ryegrass has a quick establishment rate that offers cover to the slower-establishing Kentucky bluegrass. Many perennial ryegrass **varieties** contain fungal organisms called endophytes, which increases resistance to certain turf insect pests.

For shady situations, fine fescue (*Festuca rubra* ssp) is the recommended turfgrass. Tall fescue (*Festuca arundinacea*), with its coarse leaf texture, is best for utility turf and low maintenance turf. "Dwarf turf-type tall fescues" are also finding use in lawns and sports turf. Like perennial ryegrass, several varieties also contain endophytes for better insect resistance. However, care must be given when using these endophyte-enhanced grasses near pastureland as grasses containing endophytes cause problems with grazing farm animals.

Creeping bentgrass (*Agrostis palustris*) is the turfgrass of choice for golf courses and high-quality lawns. Several other bluegrasses, fescues, ryegrasses, and bentgrasses are used to varying degrees in turf situations. Annual bluegrass (*Poa annua*) is a very tenacious bluegrass that is more often a persistent weed.

Common cool-season turfgrasses are listed in Table 3-1.

Warm-Season Turfgrasses

Of the warm-season grasses, various hybrid bermudagrasses (*Cynodon dactylon*) are best for a wide variety of turf situations in the lower transition zone and warm-season zone. Depending on the variety, bermudagrasses can be found in home lawns, sports turf, golf courses, commercial lawns, and utility turf.

TABLE 3-1 Common Cool-Season Turfgrasses

	Creeping Bentgrass *Agrostis palustris*	Kentucky Bluegrass *Poa pratensis*	Perennial Ryegrass *Lolium perenne*	Fine Fescue *Festuca rubra* ssp	Tall Fescue *Festuca arundinacea*
Growth Habit	Strong stolons	Strong Rhizomes	Bunch type	Bunch type or short rhizomes	Bunch type
Leaf Texture	Fine	Medium-fine to medium	Fine to medium	Very fine	Coarse to medium
Shoot Density	High	Medium to High	Medium	High	Low to Medium

TABLE 3-2 Common Warm-Season Turfgrasses

	Bermudagrass ***Cynodon dactylon***	St. Augustine ***Stenotaphrum secundatum***	Buffalograss ***Buchloe dactyloides***	Centipedegrass ***Eremochloa ophiuroides***	Zoysiagrass ***Zoysia* sp.**
Growth Habit	Stolons and rhizomes	Stolons	Well-developed stolons	Short stolons	Stolons and rhizomes
Leaf Texture	Fine to medium	Coarse	Fine	Medium	Fine to medium
Shoot Density	High	Medium	Medium	Medium	Medium to high

Zoysiagrass (*Zoysia* sp) is found in lawns of homeowners desiring a slow-growing turf with minimal care. It is the most cold tolerant of the warm-season grasses. Zoysiagrass will survive winters in such areas as Ohio, Pennsylvania, and southern Michigan.

St. Augustine (*Stenotaphrum secundatum*) is the turfgrass of choice for shady situations in the more humid and warmer areas of the warm-season zone. Centipedegrass, bahiagrass, and other warm-season grasses are used primarily for low-maintenance utility turf situations.

Common warm-season turfgrasses are listed in Table 3-2.

ACTIVITIES

Examination of Seed Structures

Using a hand magnifying lens or dissecting scope, locate the lemma, palea, rachilla, and awn seed structures of the sample seed material provided by your instructor. Carefully examine these structures, noting the differences in size, color, and other features. Sketch three perspectives, front, back, and side, in the space provided in the seed structures table (Figure 3-5). Label the structures in your sketches. When in doubt, ask your instructor for assistance.

	Sample 1	Sample 2	Sample 3
Grass			
Front			
Back			
Side			

FIGURE 3-5 Seed structures table

Examination of Vegetative Structures

Examine the various structures found on three different grass plants. Use the larger field grasses initially for ease of finding structures. For each structure, note and record the origin, size, color, and surface characteristics (that is, presence of hairs, rough edges, etc.). In the vegetative plant structures table (Figure 3-6), make drawings and notes as extensive and detailed as possible. The more time spent studying anatomy now will pay off later when required to identify turfgrass plants. As an out-of-lab exercise, attempt to identify the different anatomical parts on the turfgrasses found in various turf areas.

	Grass Plant 1	Grass Plant 2	Grass Plant 3
Tiller			
Rhizome			
Stolon			
Leaf			
Leaf Blade			
Leaf Sheath			

FIGURE 3-6 Vegetative plant structures table

Turfgrasses **25**

	Grass Plant 1	Grass Plant 2	Grass Plant 3
Collar			
Ligule			
Auricles			
Crown			
Roots			
Inflorescence (Flower Head)			

FIGURE 3-6 Vegetative plant structures table *(continued)*

Study of Known Turf Seed

Equipment.
- hand lenses
- cellophane or clear plastic tape
- sample seeds

Obtain and study seed samples of each species of turfgrass identified in the seed samples tables (Figures 3-7 and 3-8). Note the characteristics of various identified turfgrass seeds. Observe numerous seeds to develop a general reference. Tape the samples in the appropriate spaces in Figures 3-7 and 3-8 for later reference.

To tape seeds in the spaces:

1. Unwind two inches of transparent tape.
2. With the adhesive side facing out, stick one-half inch of one tape end to your middle finger and one-half inch of the other tape end to your thumb.
3. Use your index finger to push the tape outward to a "U" shape.
4. Dip the tape into a seed sample to cover the middle inch of tape with seed.
5. Affix the tape with seed to the spaces provided in Figures 3-7 and 3-8.

Creeping Bentgrass	Kentucky Bluegrass	Perennial Ryegrass	Fine Fescue	Tall Fescue

FIGURE 3-7 Cool-season grass seed samples table

Bermudagrass	St. Augustine	Buffalograss	Centipedegrass	Zoysiagrass

FIGURE 3-8 Warm-season grass seed samples table

Determining Seed Mixture Contents

Using the tables created in the previous activity, identify the turfgrass species in the various seed mixtures. Each mixture will have two to five species of turfgrass seed. Write your answers in the seed mixtures table (Figure 3-9), and consult with your instructor as to the correct answers. Tape samples in the spaces provided in the same manner as described in the previous activity.

Identification of Turfgrasses

Study the various laboratory samples of different turfgrasses. Use the vegetative identification key (Table 3-3) and previous structure diagrams to note the various characteristics of the different turfgrasses.

Using experience gained in the laboratory, identify the different grasses found in various assigned turf areas around the campus. Fill in the turfgrass identification table (Figure 3-10) and consult with your instructor as to the correct answers.

Mixture One	Mixture Two	Mixture Three	Mixture Four
1.	1.	1.	1.
2.	2.	2.	2.
3.	3.	3.	3.
4.	4.	4.	4.
5.	5.	5.	5.
Seed Mix Sample	Seed Mix Sample	Seed Mix Sample	Seed Mix Sample

FIGURE 3-9 Seed mixtures table

Turf Stand One	Turf Stand Two	Turf Stand Three	Turf Stand Four
1.	1.	1.	1.
2.	2.	2.	2.
3.	3.	3.	3.
4.	4.	4.	4.
5.	5.	5.	5.

FIGURE 3-10 Turfgrass identification table

TABLE 3-3 Vegetative Identification Key

A vegetative key works by deciding at each level which statement best fits the plant. If the first descriptive statement closely matches the plant, the next level below is studied. If the first statement does match, the second statement in the current level should match the plant.

Key Description Grass Species	Grass Species
I. Leaves folded in the bud	
A. Auricles present, small to long and clawlike; ligule a short membrane; veins of blade prominent; blades shiny on underside, dull on upper surface, 2–5 mm wide; bunch-type growth	**Perennial ryegrass**
B. Auricles absent	
1. Creeping stolons present	
a) Blades narrowed to form a short stalk at base of the blade; sheaths greatly compressed	
(1) Ligule a fringe of very short hairs; blade 4–10 mm wide, blunt and rounded at tips; color smooth without hairs	**St. Augustinegrass**
(2) Ligule membranous with short hairs at top; collar has hairs; blade 3–5 mm wide, edges hairy towards base	**Centipedegrass**
b) Blades not constricted; sheaths compressed	
(1) Ligule a fringe of hairs	
(a) Collar broad, hairy; blades and sheaths very hairy; blades 4–5 mm wide, V-shaped; stolons and rhizomes present	**Kikuyugrass**
(b) Collar narrow, slightly hairy; sheaths and blades smooth or slightly hairy	
(i) Blades 1.5–4 mm wide; stolons and rhizomes present; blade tip tapers to a point	**Bermudagrass**
(ii) Blades 4–10 mm wide, blunt and rounded at tip; rhizomes absent; nodes hairy	**Carpetgrass**
(2) Ligule very short, membranous; blades 4–8 mm wide, slightly hairy toward the base; stolons and rhizomes short and thick	**Bahiagrass**
2. Stolons absent	
a) Blades narrow, curled, bristlelike, prominent veins on upper surface	**Red fescue**
(1) Rhizomes present; culms reddish at base; ligule membranous and medium short; blades smooth	
(2) Rhizomes absent	
(a) Leaves-bluish green, 0.5–1.5 mm wide; culms green or pinkish at base; ligule membranous and very short; sheath split	**Sheep fescue**
(b) Leaves bright green, 1–2.5 mm wide; culms red at base; sheath closed almost to top; very similar to red fescue except for absence of rhizomes	**Chewings fescue**
b) Blades flat to V-shaped; veins inconspicuous	
(1) Blades with boat-shaped tip; transparent lines on either side of the midvein (hold up to light to see this)	
(a) Rhizomes normally absent	
(i) Blades usually light green; smooth sheath whitish at base; membranous ligule long and pointed; seed heads often present	**Annual bluegrass**
(ii) Thin stolons present; sheaths have rough, bumpy feel; membranous ligule long and pointed	**Rough bluegrass**

TABLE 3-3 Vegetative Identification Key *(continued)*

Key Description Grass Species	Grass Species
(b) Rhizomes present	
(i) Blade tapers to boat-shaped tip; sheaths strongly compressed; membranous ligule is medium short	Canada bluegrass
(ii) Blade does not taper to a point, entire leaf has similar width; sheaths not strongly compressed; very short membranous ligule	Kentucky bluegrass
(2) Blades without a boat-shaped tip; no transparent lines on either side of midvein; prostrate growth habit	Goosegrass
II. Leaves rolled in the bud	
A. Auricles present	
1. Sheaths reddish at base; blades shiny on underside	
a) Edges of blades smooth, leaf width 3–7 mm; long clawlike auricles	Annual ryegrass
b) Edges of blades rough; short auricles	
(1) Auricles usually without hairs	Meadow fescue
(2) Auricles and collars have a few short hairs, auricles very small	Tall fescue
2. Sheaths not reddish at base; underside of leaves not shiny	
a) Strong rhizomes present; long clasping auricles	Quackgrass
b) Rhizomes absent; long clawlike auricles	Crested wheatgrass
B. Auricles absent	
1. Sheaths round	
a) Collar somewhat hairy	
(1) Sheaths not hairy	
(a) Strong stolons present	
(i) Collar with long hairs; ligule a fringe of hairs; blades 2–5 mm wide; rhizomes	Japanese lawngrass
(ii) Collar sparsely hairy; blades 2–3 mm wide; similar to Japanese lawngrass	Manilagrass
(b) Stolons not present; weak rhizomes; collar has long hairs; leaf width 1–2 mm	Blue gamma
(2) Sheaths not hairy; no rhizomes	Downy bromegrass
b) Collar not hairy	
(1) Ligule a fringe of hairs; collar board	
(a) Rhizomes present; blade smooth	Mascarenegrass
(b) Rhizomes absent, stolons present; blade hairy, 1–3 mm wide, grayish-green	Buffalograss
(2) Ligule membranous; collar narrow	
(a) Sheath closed almost to the top; blade 8–12 mm wide, smooth Smooth bromegrass	Smooth bromegrass
(b) Sheaths split with overlapping edges; ligule pointed; blades with prominent veins on upper surface	
(i) Stolons absent or weak	
(a) Blades 3–7 mm wide; long ligule	Redtop
(b) Blades 1–3 mm wide; ligule medium-short	Colonial bentgrass
(ii) Strong stolons	
(a) Blades 1 mm wide, ligule medium-short	Velvet bentgrass
(b) Blades 2–3 mm wide; ligule medium length; long stolons	Creeping bentgrass
2. Sheaths compressed (flattened); ligule very short, membranous; blades 4–8 mm wide; stolons and rhizomes short and thick	Bahiagrass

EXERCISES

Turfgrass Plant Collection and File

Collect, identify, and mount on herbarium sheets, or turfgrass plant collection cards (Figure 3-11), four turfgrass species from maintained turf areas. The turfgrass species should be selected from the list of turfgrasses discussed earlier in the unit. Use the vegetative identification key (Table 3-3) information gathered in previous activities, and other descriptions for proper identification.

Procedure.

1. Select 5-10 single plants (not plugs) identified and found in a *maintained* turf area. Plants may be removed from the soil using a variety of digging implements, including a knife or trowel. Record pertinent data at the time of collection, including date, location (city, state, county), collector name, locality (lawn, sports field, golf course), and habitat (full sun, shade, damp soils). When collecting creeping bentgrass or hybrid bermudagrass, do not take specimens from a golf green. Collect these specimens from the apron only.
2. Place the collected specimens within single folds of newspaper. Arrange the leaves of the specimens so that both the front and back sides of the leaves can be observed. All soil must be removed from the roots. Washing may be necessary to remove soil.
3. Place the specimens, now within the newspaper, between two pieces of blotter paper or corrugated cardboard, and then place the whole arrangement under heavy weights such as several textbooks. (Do not use the turf textbook.)
4. Drying should occur within 3-8 days. Check daily to ensure mold and mildew do not start growing on your specimens. Change the blotter paper or cardboard as necessary for drying. Do not try

Tech. Name: ____________________
Com. Name: ____________________
City: ______________ State: _____
Local: ____________________
Collector: ____________________

FIGURE 3-11 Herbarium sheet (turfgrass plant collection card). Duplicate this card four times to complete the exercise.

the following methods: hair dryers, heating vents, conventional ovens, or microwave ovens. None of these methods works effectively.
5. When dry, select 2-4 single plants for mounting. Showing front and back arrangements, mount the specimens in the spaces provided on the herbarium sheets (Figure 3-11) with paste, muscilage, linen tape, white glue, or transparent adhesive film. Do not use common cellophane tape, transparent tape, rubber cement, epoxy, or plastic adhesives (super glues).
6. Label each card in the lower right corner with the appropriate information. Type or print neatly with black ink *before* mounting specimens.
7. Turn in the sheets for evaluation on the date provided by your instructor.

Using Paper. Place your herbarium sheets into your three-ring binder for later reference. Develop a form to maintain data on different species and cultivars.

Using a Computer. Develop a computer database of the different turfgrasses (Figure 3-12). If a scanner is available and the database software supports graphics, scan your specimens into a graphics file for placement in the data file.

On-Line Companion. U03E01.WDB—Turfgrass species database

Turfgrass Cultivar Information File

Develop a file of information on the different turfgrass cultivars grown by the various turfgrass producers. Collect advertisements of the different turfgrass cultivars. Photocopies of the advertisements are acceptable.

Using Paper. Develop a form for recording and organizing the information on the various turfgrass cultivars. Keep this completed form along with the advertisements in your three-ring binder just behind the turfgrass plant collection cards.

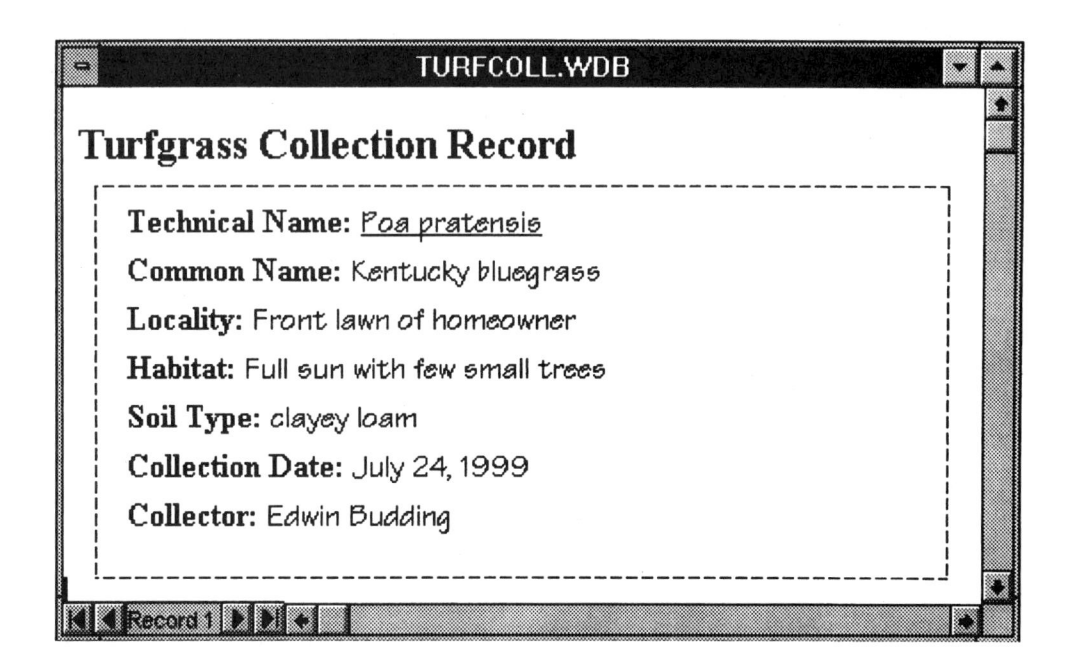

FIGURE 3-12 Sample turfgrass data form

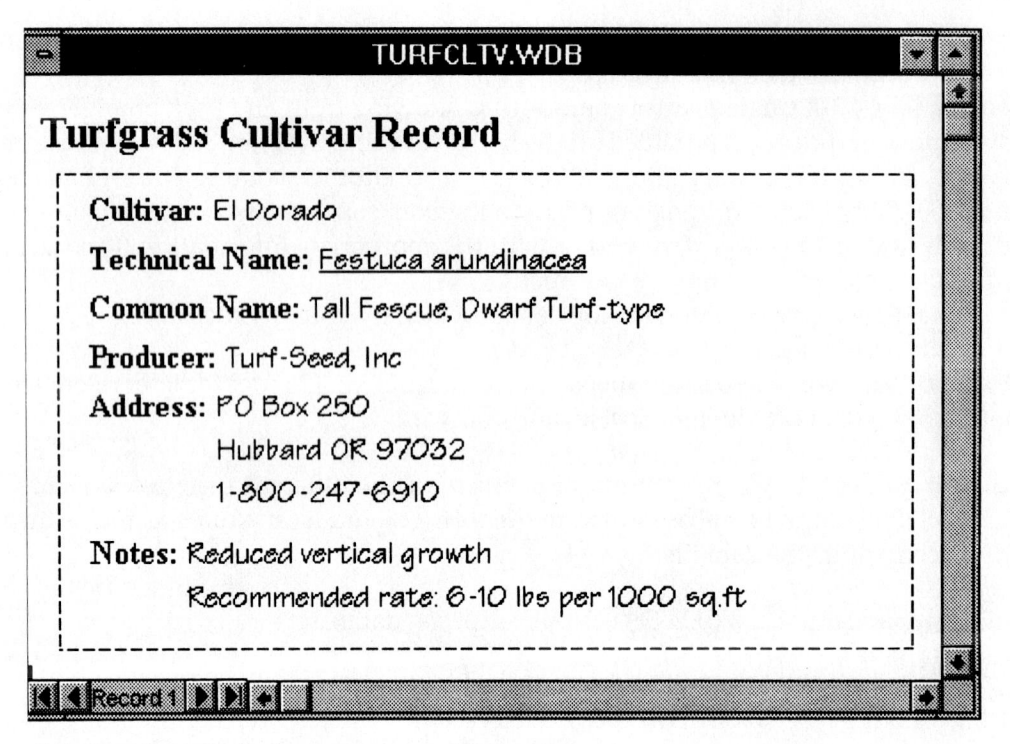

```
┌─────────────────────────────────────────────────┐
│  −                    TURFCLTV.WDB          ▼ ▲ │
├─────────────────────────────────────────────────┤
│  Turfgrass Cultivar Record                      │
│  ┌───────────────────────────────────────────┐  │
│  │ Cultivar: El Dorado                       │  │
│  │ Technical Name: Festuca arundinacea       │  │
│  │ Common Name: Tall Fescue, Dwarf Turf-type │  │
│  │ Producer: Turf-Seed, Inc                  │  │
│  │ Address: PO Box 250                       │  │
│  │          Hubbard OR 97032                 │  │
│  │          1-800-247-6910                   │  │
│  │                                           │  │
│  │ Notes: Reduced vertical growth            │  │
│  │        Recommended rate: 6-10 lbs per 1000 sq.ft │
│  └───────────────────────────────────────────┘  │
│  |◄ ◄ Record 1 ► ►|                              │
└─────────────────────────────────────────────────┘
```

FIGURE 3-13 Sample turfgrass cultivar record

Using a Computer. Develop a database file of the different turfgrass cultivars (Figure 3-13). Include information on characteristics distinguishing the cultivar from other cultivars. If a scanner is available and the database software supports graphics, scan the advertisements into a graphics file for placement in the data file; otherwise, keep the advertisements in your three-ring binder. For advanced computer users who are using relational database managers, a separate file of cultivar information can be developed with a relational index key to the turfgrass plant collection data file.

On-Line Companion. U03E02.WDB—Turfgrass cultivar file

UNIT 4

Turf Soils

The **soil** is an integral part of the growth and survival of turfgrass. It is the main source of water and minerals for the turfgrass plant. It is also a means of achorage for the turfgrass plant. An often neglected part of turfgrass management, many turf problems are avoidable through an understanding of soils and a mangement of the soil beneath the turfgrass.

There are many different definitions for soil. To a geologist, soil is highly weathered rock. For turf managers, the best definition of soil is a dynamic body of inorganic and organic material with the potential to support life. This includes not only turfgrass plants but any plant or animal life. In a turf community, nongrass plants and all animals living in the soil have an impact on turfgrass growth. Earthworms are an example of a soil animal that can greatly enhance turfgrass growth.

There are many different properties attributed to soils. These properties consist of physical as well as chemical properties. Soil **organic matter** is a unifying component between the physical and chemical properties of soils.

TURF SOIL PHYSICAL PROPERTIES

Soil physical properties are those properties measurable by physical means. These properties include the size and shape of soil particles, the relative proportions and arrangements of the particles, and the relationship of water and air to the solid component of soil.

Soil Texture

Of all the properties of soil *soil texture* probably has the greatest impact on turfgrass management in terms of plant growth, water management, and fertility management. **Soil texture** is the relative proportions of *soil separates,* which are classes of different-sized soil particles.

The three major groups of soil separates are *sand, silt,* and *clay.* Sands are the largest of the soil particles, with a maximum size of 2.00 mm. Sands are followed in size by silts, at 0.05 mm. Clays are the smallest soil particles, at less than 0.002 mm in size (Table 4-1).

TABLE 4-1 USDA Soil Separate Classification

Separate Name	Diameter (millimeters)	Particle Size Analogy
Very coarse sand	2.00–1.00	8 ft diameter ball
Coarse sand	1.00–0.50	4 ft diameter ball
Medium sand	0.50–0.25	2 ft diameter ball
Fine sand	0.25–0.10	Basketball
Very fine sand	0.10–0.05	Softball
Silt	0.05–0.002	Golf ball
Clay	Less than 0.002	Popcorn kernel

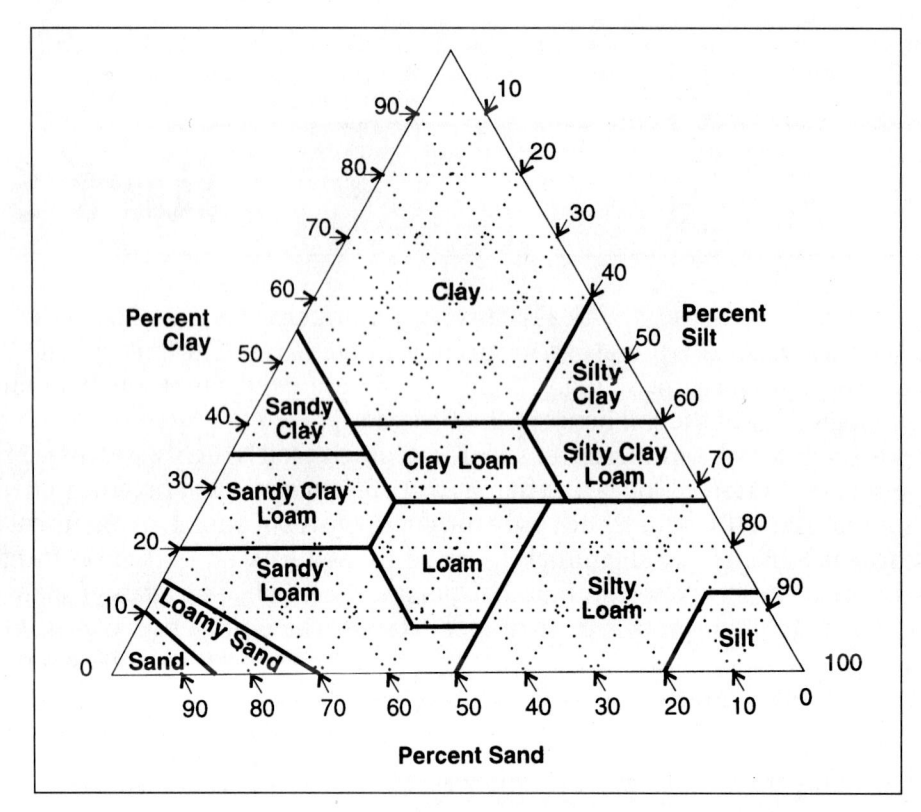

FIGURE 4-1 Soil texture triangle

Soil texture encompasses the *soil texture classes,* which are the relative proportions of the three major soils separates. These classes range from coarse-textured soils (composed primarily of sand, with little silt and clay) to fine-textured soils (which are high in clay and silt with little sand). A medium-textured soil, or loam, contains about 40 percent sand, 40 percent silt, and 20 percent clay. The soil texture triangle (Figure 4-1) is a common means of determining and displaying the soil texture classes.

Soils high in sand generally will have poor water-holding characteristics, poor "mineral nutrient" retention characteristics, but good water-drainage and **aeration** characteristics. Sandy soils also have a limited tendency to compact, which is a desirable characteristic for sports fields and other high-traffic turf areas.

Soils high in clay have many of the opposite characteristics to a sandy soil. Clays will compact easily, leading to more pronounced water movement problems and reduced aeration. Many types of clays can hold certain mineral nutrients for controlled release to a plant. To overcome problems associated with clay soil, turf managers can, through development of good soil structure, modify the soil so that it exhibits characteristics similar to those of medium-textured soils.

Soil Structure

Soil structure is the arrangement of soil particles into larger aggregates, or clumps known as *peds*. With the development of soil structure, there is the creation of larger pores, known as *macropores*. In comparison to the smaller *micropores,* which are important in water retention, these macropores allow for better water movement through the soil as well as better aeration and plant root growth. Although there are several different classes of soil structure, the two types of soil structure of major concern to the turf manager are granular soil structure and "structureless" soil structure.

A granular soil structure consists of spherical-shaped peds 1-10 mm in diameter. This structure produces the most optimum balance of macropores and micropores necessary for water management, soil aeration, and plant growth. A crumb structure is similar to a granular structure except the peds are less dense and stable.

A "structureless" soil structure means little or no structure. Pure sand soils have no structure, consisting primarily of individual particles and little or no peds. A compacted soil is essentially a soil having very little structure because of traffic breaking and compressing most of the peds into a large massive structure.

Structure development is the result of climate, soil organisms, and in the case of turf, people acting on the soil. Expansion and contraction of soils through freezing, thawing, wetting, and drying along with soil organisms working through a soil (e.g., earthworms, plant roots, and **moles**) are all instrumental in developing soil structure. Tillage of agricultural land and core **aerification** of turf are mechanical methods used to develop structure.

Organic Matter

Soil organic matter provides the "glue" to hold soil particles together into larger aggregates. Good soil structure is hard to maintain in soils low or devoid of organic matter. Cultivation of soils under an established turf is not as easy to perform as cultivation of soils in an agronomic crop field. The turf manager should strive to increase organic matter content to maintain good soil structure.

Water in Turf Soils

The main source of water for turfgrass plants is the soil. Although turfgrass leaves can absorb small amounts of water, most of the water obtained by the plant comes from the soil. Soil is similar to a storage tank that holds water between periods of rainfall or **irrigation**. The turf manager needs to be sure water is moving into this tank and not "leaking" back out too quickly. The soil should hold enough water to reduce irrigation need but not hold so much water as to minimize aeration to the point that turfgrass roots suffer from oxygen depletion.

Turf Soil Water Infiltration and Drainge. Water moves into a soil due to *infiltration*. Good infiltration ensures that the soil contains enough water to meet plant needs. Rapid infiltration is also important on sports turf for the removal of casual water (i.e., a temporary ponding situation). Besides interfering with play, casual water contributes to **compaction**, puddling of soil, and sun scald of turfgrass plants. Improper use of the turf (such as the high traffic generated by a band concert on a football field) can quickly destroy good infiltration by way of compaction.

Soil texture, soil structure, degree of compaction, **thatch**, and soil chemical conditions all have an impact on soil water infiltration. Through proper (or improper) cultural or management practices, a turf manager can increase or reduce soil water infiltration.

An infiltrameter is a device for determining the infiltration rate of a soil. This device consists of one ring or a set of double concentric rings driven vertically into the soil. The soil within the rings is thoroughly saturated with water. The rate of disappearance of a known volume of water in the inner ring is divided by the surface area of the inner ring to give an infiltration rate in centimeters per hour (cm/hr). Due to lateral movement of water, single-ring infiltrameters are not as effective in determining infiltration rates as are double-ring infiltrameters. For the turf manager, single-ring infiltrameters made from common materials are acceptable for determining infiltration rates of a turf area.

Excessive amounts of water move through the macropores of a soil as drainage. Poor drainage leads to excessive amounts of water in the soil. This, in turn, leads to root loss due to lack of adequate oxygen. In addition, excessive water leads to a greater tendency for soil compaction, which, in turn, compounds problems of poor drainage and poor aeration.

In turf, both surface and subsurface drainage are important in maintaining proper soil water and aeration levels. For many turf areas, particularly sports fields, a turf manager will install additional, artificial drainage by installing an underground drainage tile. Along with the addition of sands, this helps to minimize compaction and disruption of soil water movement.

Water Retention and Movement. Micropores retain water in the soil through the unique nature of water molecules to "cling" to surfaces and other water molecules. Soil water movement that is important to plant growth occurs through the "capillary" movement of water, which works due to the adhesive and cohesive nature of water.

Due to small particle size and a plate-like particle shape, clay soils tend to have a large number of micropores, which results in good water-retention capability. Micropores will hold some water so tightly however, that the water is unavailable to most turfgrass plants.

In sands, because of the large particle size and sphere-like particle shape, there are not enough micropores to hold adequate amounts of water. With the addition of organic matter to act as "sponges," the water retention capabilities of sands can be increased to some degree.

TURF SOIL CHEMISTRY AND FERTILITY

The soil chemical properties important to a turf manager are those involving the supply, availability, and retention of inorganic nutrients necessary for plant growth and survival.

Nutrients for Plant Growth

All plants need sixteen chemical elements, or nutrients, for existence and growth (Table 4-2). Soil provides thirteen of these sixteen elements. Many of these elements come from the decomposition of minerals during continual soil formation. For example, potassium comes from the weathering and breakdown of feldspar minerals found in igneous rock from volcanic action. Other elements

TABLE 4-2 The Sixteen Elements Necessary for Plant Growth

Element	Chemical Symbol*	Chemical Forms Absorbed by Plant
Hydrogen	H	H_2O and others
Oxygen	O	CO_2, H_2O
Phosphorus	P	HPO_4^{--}, $H_2PO_4^{-}$
Potassium	K	K^+
Nitrogen	N	NO_3^-, NH_4^+
Sulfur	S	SO_4^{--}
Calcium	Ca	Ca^{++}
Iron	Fe	Fe^{++}, Fe^{+++}
Carbon	C	CO_2
Boron	B	$H_2BO_3^-$
Magnesium	Mg	Mg^{++}
Chlorine	Cl	Cl^-
Molybdenum	Mo	MoO_4^{--}
Manganese	Mn	Mn^{++}
Copper	Cu	Cu^{++}
Zinc	Zn	Zn^{++}

* As a learning aid, the chemical symbol column is arranged as a mnemonic. Going down the column, it reads "**HOPKiNS CaFe, C. B. M**a**n**ager. **Cl**osed **Mo**nday **M**ornings, **C**eeyo**u Z**oo**n**!"

are converted by various means into compounds useable by the plant. Nitrogen is a gaseous element converted into ammonias and nitrates for uptake by plants.

The elements are classified into groups depending on the amount needed by the plant. Plants require *macronutrients* of nitrogen, phosphorus, potassium, magnesium, calcium, and sulfur in larger amounts than the **micronutrients** of iron, boron, manganese, copper, zinc, and molybdenum and chlorine. Calcium, magnesium, and sulfur often are classified as *secondary nutrients* because plants require a lesser amount of these nutrients than they do nitrogen, phosphorus, and potassium. For turf, the turfgrasses often need nitrogen in the greatest amounts, followed by potassium and phosphorus.

Soil Fertility

Soil fertility refers to the amount and availability of the elemental nutrients in the soil. It should be noted that just because a soil is fertile does not necessarily mean the soil will provide nutrients for plant growth. In many soils, nutrients are unavailable due to chemical binding of the nutrient to soil particles. Productive soils are those soils where nutrients are dissolved in soil water or temporarily tied up on soil particles.

All the elements are found in the soil in ionic form as either positively charged ions (**cations**) or negatively charged ions (*anions*). Many soils, particularly those high in clay or organic matter, can temporarily hold cation-based nutrients for later release to the plant. This is known as the **cation exchange capacity** or **CEC** of a soil. As with water, the soil serves as a storage tank to reduce the need for continual applications of additional nutrients. Cation exchange can serve as an indicator of the capacity of the storage tank.

Anions, unlike cations, rarely attach to soil particles. Anions will remain in soil solution, which allows them to leach out of reach of turf roots through excessive water drainage. Nitrogen is often in an anionic form in the soil and, therefore, easily lost through leaching. Proper irrigation is important in helping to maintain adequate nitrogen levels in the soil.

Soil Reaction

In the process of cation exchange and other chemical reactions, a soil will develop varying concentrations of hydrogen cations (H^+) and hydroxyl anions (OH^-). These concentrations determine the **pH** of the soil. **Acidic soils** are high in hydrogen cations and low in pH (with a pH number below 7). **Alkaline soils** are low in hydrogen cations and high in pH (with a pH number above 7). Neutral soils have an even balance of hydrogen and hydroxyl ions, with a pH number of 7.

Soil pH indirectly affects plants. Nutrients are readily available in varying concentrations depending on the pH of the soil (Figure 4-2). At certain pH levels, certain nutrients react with other chemicals and become unavailable to the plant. Adjustment of the pH will often result in the release of the nutrient back into soil solution.

SOIL TESTING AND RECORDS

Because soil serves as a storage medium for both water and nutrients, a turf manager should maintain records of the various physical and chemical properties of the soils underlying a turf area. Knowledge of various physical properties, such as texture and **bulk density**, provides the turf manager with key information in developing an efficient irrigation program. Various chemical properties and nutrient levels combined with information on physical properties aid in implementing a cost-effective fertilization program.

Testing of various soil chemical properties and soil fertility levels is common practice. Turf managers of high-maintenance sports turf (such as golf courses), however, should also determine and

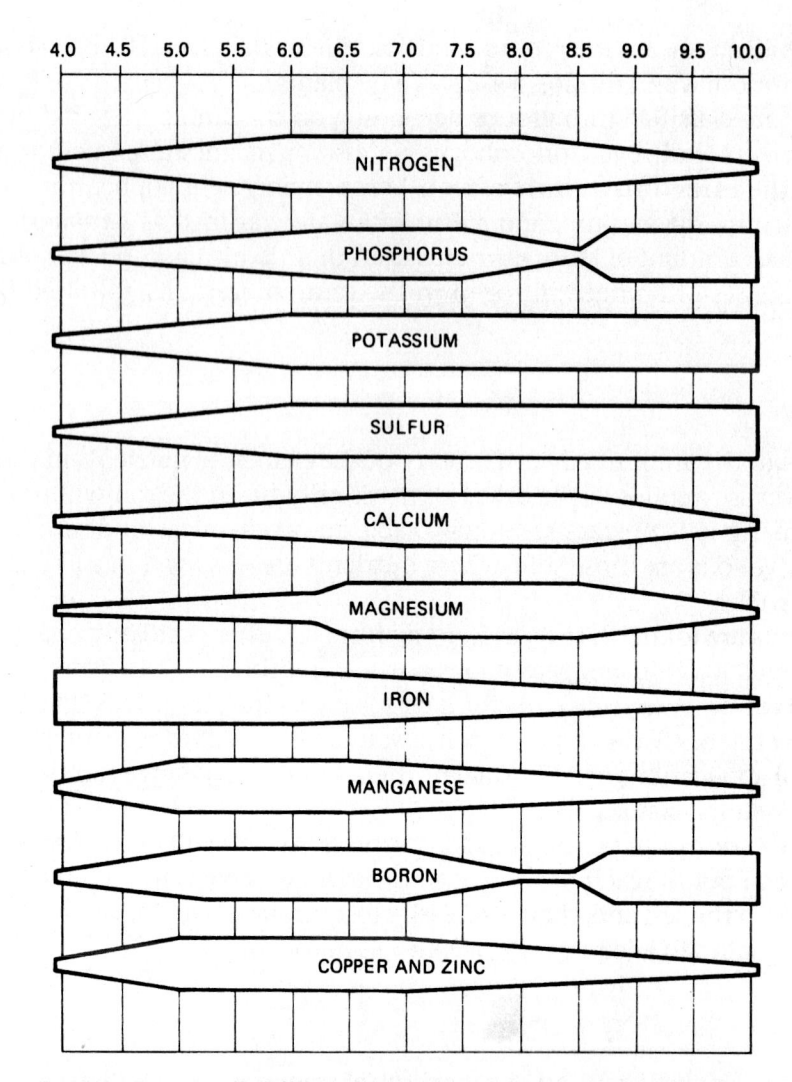

FIGURE 4-2 Availability of nutrients at various pH levels

maintain records on the various physical properties of the soil or soils underlying a turf. These properties include texture, bulk density, water-infiltration rates, **water-percolation** rates, and compaction tendency. The turf manager can perform many of these tests in the field with minimal effort and equipment.

Soil testing for chemical properties such as pH and fertility levels can be done in the field using colormetric soil test kits. With some practice, the turf manager can obtain acceptable results. The chance for error, however, is much greater because of factors such as contamination or breakdown of testing solution. For more accurate results, the turf manager should send soil samples to a soil testing lab for analysis and comparison.

Regardless of the testing methods, the turf manager should repeat soil tests on a routine basis—usually every one to three years, depending on the use of the turf and other cultural practices. (For example, an irrigated golf green will need testing more often than will a moderately maintained home lawn.)

Because soil tests will be performed on a routine basis, the turf manager will need to update records to reflect any changes in the soil. With accurate, updated records, the turf manager can adjust various cultural practices, such as the fertilization program, to reflect current conditions in the soil.

ACTIVITIES

Soil Texture Analysis Using the Quart Jar Method (Figure 4-3)

Equipment.

- one-quart canning or mayonnaise jar with lid
- one-half cup (approximately) of soil
- automatic dishwasher detergent or table salt
- water
- ruler with metric divisions

Procedure.

1. Place soil in jar with one heaping teaspoon of detergent or salt and fill with 3-1/2 cups of water.
2. Cap and shake for 5 minutes, then place jar in an undisturbed location for 24 hours.
3. After 24 hours, measure the depth of settled soil.
4. Shake thoroughly for 5 minutes.
5. Gently place jar on table and let it stand for 40 seconds, then quickly measure the depth of settled sand.
6. Wait 2 hours and measure depth of settled soil (sand and silt). Subtract this depth from depth of sand (obtained in step 5) to get the depth of silt layer. The remaining unsettled depth is clay.
7. Convert to percentage of sand, silt, and clay is displayed in Table 4-3, then use the soil texture triangle (Figure 4-1) to determine the textural class.

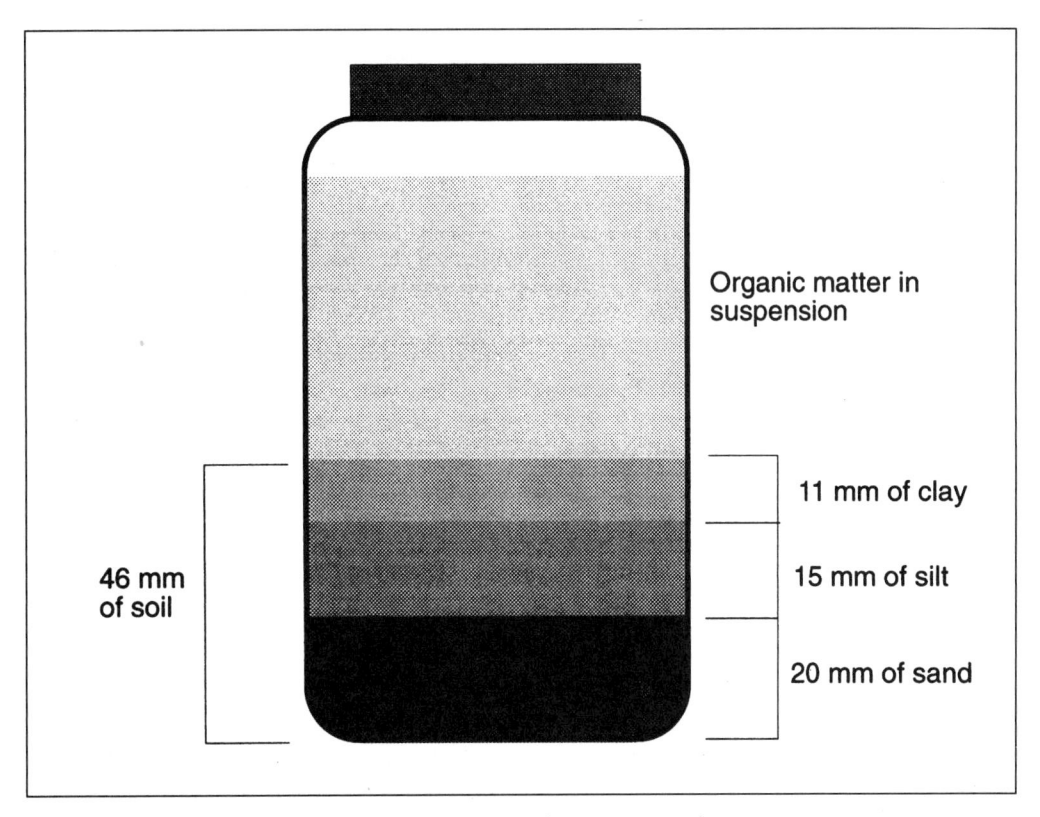

FIGURE 4-3 Determining soil texture using a quart jar

TABLE 4-3 Calculating Soil Texture Percentages

$\dfrac{\text{sand depth}}{\text{total depth}} \times 100\% = \%$ sand	$\dfrac{20 \text{ mm}}{46 \text{ mm}} \times 100\% = 43\%$ sand
$\dfrac{\text{silt depth}}{\text{total depth}} \times 100\% = \%$ silt	$\dfrac{15 \text{ mm}}{46 \text{ mm}} \times 100\% = 33\%$ silt
$\dfrac{\text{clay depth}}{\text{total depth}} \times 100\% = \%$ clay	$\dfrac{11 \text{ mm}}{46 \text{ mm}} \times 100\% = 24\%$ clay

Measuring Soil Infiltration with an Infiltrameter

Determine soil infiltration rates in assigned locations of campus turf areas. Use a single ring infiltrameter as described following. Fill in the infiltration rates table (Figure 4-4) describing the tested area (bare soil, turf covered, compacted, sand, clay, etc.). Determine the infiltration rates in centimeters per hour (cm/hr).

Equipment.

- tin cans open at both ends, sections of beveled pipe, or golf-green putting cups
- ruler
- graduated beaker or cylinder
- mallet or hammer
- block of wood

Procedure.

1. Measure the diameter of the provided tin can, pipe section, or putting cup, and calculate the surface area (in square centimeters [cm^2]) of one end.
2. Place one end of the cylinder over a selected test area and carefully drive it 1–2 inches into the ground. If using a golf-green putting cup, turn it upside down.
3. Fill the infiltrameter with water and allow it to drain into the soil for pre-wetting. (Be careful when pre-wetting clay; do not add a large amount of water.) Also drench the area surrounding the infiltrameter.

	Test Area 1	Test Area 2	Test Area 3	Test Area 4	Test Area 5
Location					
Description					
Soil Texture (Estimate by feel)					
Infiltration Rate (cm/hr)					

FIGURE 4-4 Infiltration rates of tested areas table

4. Fill the ring with a specified amount of water and measure the amount of time required for it to disappear beneath the soil surface.
5. Calculate the infiltration rate for the soil using the following formula:

$$\textbf{Infiltration rate} = \frac{(\text{amount in cm}^3 / \text{cm}^2 \text{ of surface area})}{\text{time in hours}}$$

where 1 milliliter of water is approximately equal to 1 cubic centimeter.
6. If you anticipate a very slow rate, such as with a clay or compacted soil, measure the depth of the water when you initially fill the cylinder with water. After a length of time, measure the depth of water remaining. For example, if the initial depth is 10 cm and after 1 hour the depth is 5 cm, calculate the infiltration rate directly as 5 cm/hr.

Soil Testing by a Soil Testing Lab

With assistance from your instructor, collect and prepare soil samples from several campus turf areas for submission to a soil testing lab. Record all pertinent information as required by the lab submission form (Figure 4-5). Collect separate samples from diverse turf areas such as lawns, sports fields, and utility turf areas. If a large turf area has areas of significantly diverse soils, separately sample each area. Save all extra soils for a later lab activity.

When the soil test lab returns the results, enter the basic data into the basic soil test data table (Figure 4-6, page 44) for use in a later activity.

Procedure.
1. Obtain the appropriate collection bags and forms from the soil testing lab.
2. Collect up to ten soil samples from each turf area. If the area contains diverse soils, collect separate samples. Each sample should be approximately 1 inch (2.54 cm) in diameter to a depth of 3-4 inches (7.6-10.2 cm).
3. Remove all surface vegetation including the thatch, unless otherwise instructed by the testing lab.
4. In a clean, plastic container or cardboard box, combine the samples of each area into a "composite" sample. Do not use plastic pails that previously held fertilizers or any metal buckets.
5. Allow the soil to air dry by spreading on clean paper.
6. Place about a cup (8 oz or 236.5 ml) of soil in the collection bag, and label with the information as directed by the soil testing lab instructions.
7. Fill out any necessary forms required by the lab before submission.

Soil Testing Lab Field Trip

Visit a university or commercial soil testing lab that will be testing the samples from the previous activity. Observe the methods of testing performed by the laboratory in determining various soil characteristics. Note other tests the lab may perform that may be of benefit to you as a turf manager (such as compost and sludge analysis).

Soil Testing with Soil Test Kits

Using the soil samples collected in the previous activity, test the soils using one or more colormetric soil test kits. Follow the directions provided in each kit. Enter the data into the basic soil test data table (Figure 4-6) and compare the results from the test kits with the results from the soil testing lab.

FIGURE 4-5 Sample soil test submission form

SAMPLING INSTRUCTIONS	FORM INSTRUCTIONS

HOW TO TAKE A GOOD SOIL SAMPLE
 	
THE KEY TO SAMPLING IS TAKING A REPRESENTATIVE SAMPLE!!
 	
1. **SAMPLE THE ROOT ZONE** - Most turfgrasses extract nutrients from the top two to three inches of soil. Therefore, soil samples for turfgrasses should be taken to a depth of **THREE INCHES**. If roots are more shallow or confined to the thatch, it is still the top three inches of soil where proper nutrients and pH should be maintained. Most ornamental plants extract nutrients from the top five to six inches of soil. Therefore, take soil samples for ornamentals to a depth of **FIVE to SIX INCHES**.
    For establishment of turf and ornamentals, a soil depth equal to the depth at which fertilizer and lime will be incorporated is the desired sampling depth.
 	
2. **TAKE A REPRESENTATIVE SAMPLE - A MINIMUM OF 10-12 CORES** (about one cup) are required for an accurate soil test. More cores should be taken on areas greater than 8,000 sq. ft. For larger areas, randomly take 12-15 samples per acre, collect cores in a clean plastic bucket, mix well and place a portion (a cup or more) in the soil sample bag.
 	
Random Lawn Samples   **Special Areas**
 	
3. **AVOID CONTAMINATION BY FERTILIZERS AND LIME** - Inaccurate soil tests results may result if samples are taken within four weeks of fertilizer or lime applications. Avoid using equipment or containers which may be contaminated with fertilizer or lime.
 	
4. **PROPERLY IDENTIFY THE SAMPLE** - Mark the soil sample bag with a positive identification. A customer name or customer number and/or place or location of the sample.
 	
5. **COMPLETE THE CLC LABS TEST REQUEST FORM** —Follow the instructions shown to the right.
 	
Please call us with any questions - **BEFORE** sampling!
(614) 888-1663
	INSTRUCTIONS FOR COMPLETING THIS FORM
	8 Simple Steps to Quick Results
	
	SAMPLE IDENTIFICATION INFORMATION:
	
	1. In the **SUBMITTED BY** block fill in your company name and address. Fill in your CLC LABS account number and a contact person's name.
	
	2. The **SUBMITTED FOR** block is only used if you are sending several samples for a single site/client. For example, if you are sending several samples from a large commercial client, enter the name of the account in this block. The SUBMITTED FOR block is not normally used when sending samples from several different customers.
	    (Your account is set up for a specific report form type. If you want the alternate report form type please call for instructions.)
	
	3. Mark the soil sample bag with the corresponding laboratory number printed in the **LABORATORY NUMBER** column. Use the red number **and** the green printed sequence number (0 thru 9), such as 4 3 2 1 - 0 .
	
	4. Fill **in your** sample identification from the bag. Your customer name or account code is suggested.
	
	FERTILIZER RECOMMENDATION INFORMATION:
	(If this section is left blank, no fertilizer recommendations will be made.)
	
	5. In the **AREA TYPE** section fill in the area type that best identifies the <u>use</u> of the area sampled.
	
	   AREA TYPES for turf: **AREA TYPES** for ornamentals:
	   LAWN, PARK, ATHLETIC FLD, BED, GARDEN,
	   CEMETERY, etc. LANDSCAPE, etc.
	
	6. Select the **PLANT CODE** that best describes the grass type or ornamental plant. Select the desired grass type in mixed grass situations.
	
	7. Select the **FERT/MAINT LEVEL** that best fits the area being sampled.
	
	   HIGH LEVEL FERTILITY/MAINTENANCE
	   Select this option for golf courses, high maintenance turfs, sod production, nursery stock production and other situations where rapid growth of ornamentals is desired.
	
	   MEDIUM LEVEL FERTILITY/MAINTENANCE
	   Select this option for home lawns and landscapes, well-maintained commercial and institutional grounds, and other situations where normal growth of turf or ornamentals is desired.
	
	   LOW LEVEL FERTILITY/MAINTENANCE
	   Select this option when maintaining low input turf and ornamentals. It should also be used for maintaining current health of mature ornamentals.
	
	   NEWLY ESTABLISHED FERTILITY
	   Select this option when sampling is done prior to establishment (planting) or during the first three to six months following establishment (planting).
	
	8. Select the desired **TEST CODE(S)** for each sample. See the CLC LABS Schedule of Fees for test descriptions.
	
	SHIPPING:
	Samples should be boxed up along with the **TOP COPY** of our form and sent via UPS or regular mail.

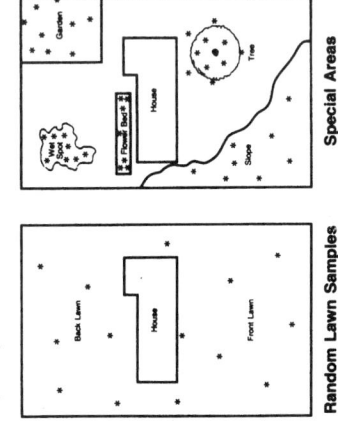

FIGURE 4-5 Sample soil test submission form *(continued)*

	Soil 1		Soil 2		Soil 3	
	Testing Lab	Test Kit	Testing Lab	Test Kit	Testing Lab	Test Kit
Nitrogen						
Phosphorus						
Potassium						
pH						
Lime Requirement						

FIGURE 4-6 Basic soil test data table

EXERCISES

Soil Test and Analysis Record File

Using information from the site inventory exercise (Unit 2), test and analyze the soils of the various turf areas of the site to obtain the following information:

- soil texture class
- infiltration rate
- information from soil lab tests

Record the findings on the site inventory. If a large turf area has significantly diverse areas of soil texture, sample and record each area separately.

Using Paper. Create an additional paper record form of soil information to be included with the site inventory records. Use the basic soil test data table (Figure 4-6) as a sample form.

Using a Computer. Modify the site inventory database to include soil analysis information. This can be accomplished by adding additional fields to the record format in the file. Use the fields in the basic soil test data table (Figure 4-6) for possible field definitions. Advanced computer users who are using relational database managers can develop a separate file of soil information with a relational key to the original site inventory file.

On-Line Companion. U04E01A.WDB–Site inventory with soil test results database
U04E02B.WKS–Site inventory with soil test results spreadsheet

U N I T 5

Turf Establishment

The key to having a quality, easy to maintain turf is proper establishment. In many situations, problems with a turf can be traced back to improper establishment. Poor selection of a turfgrass, such as creeping bentgrass for a home lawn, can result in an expensive management program that a homeowner cannot afford. Failure to put effort in developing a good soil bed could lead to problems for the life of the turf.

There are several methods of establishing turf within the two broad categories of seeding and vegetative planting. Vegetative planting consists of **sodding**, **plugging**, **stolonizing**, and **sprigging**. Seeding is the most common method of planting cool-season turfgrass, although some of the creeping-type grasses are planted as sod. Vegetative planting is the chief method of planting warm-season grasses because of the lack of usable seed supplies. The method chosen by the turf manager will depend on several factors including the species of grass, the time of season, and costs.

SELECTING A TURFGRASS

Selecting a turfgrass for a particular site depends on several major factors:

- climate
- environment
- use of the turf
- quality level
- maintenance level

Climate

Various grass species grow best within an optimum temperature range. As mentioned in Unit 3, the areas that exhibit these optimum temperature ranges are the cool-season zone and the warm-season zone.

Along with temperature, rainfall is an additional climatic factor. Within the temperature zones there are humid, semi-arid, and arid regions. The amount of rainfall within these regions has an impact on planting conditions. Grasses that do well in the humid regions will require irrigation in the more arid regions. Conversely, some dryland grasses may not grow well under humid conditions.

Environment

On a localized level, various environmental factors influence turfgrass selection. Of major concern are light conditions, in particular shade conditions. Several turfgrass species require full sun to grow and survive. A few turfgrasses, such as fine fescue (*Festuca rubra*) and St. Augustine (*Stenotaphrum secundatum*) can tolerate the lower light levels of shaded areas within a landscape.

In addition to light, other environmental factors that may impact the turf manager's selection of a turfgrass species include soil water conditions, topography, and air movement (wind).

Use

The specific use of a turf is a major factor in selecting a turfgrass. Creeping bentgrass is an appropriate and necessary choice for the golf course but may not be a reasonable selection for the home lawn because of the high maintenance requirements. To establish a utility turf that will receive little or no maintenance other than occasional mowing, the turf manager may select grasses apart from the group of grasses normally used for turf.

Quality Level

Leaf texture, density, and growth habit all have an impact on the visual quality of a turf. The desire for a smooth, carpet-like turf requires the selection of a grass that is capable of creating the desired quality level.

Cultural Level

Associated with the quality level, the turf manager must determine how much time, labor, and money are available to apply the various cultural practices necessary to maintain a turf area. As an example, many homeowners desire a lawn that requires limited mowing, low fertilizer applications, and irrigation by natural rainfall. Golf course superintendents, however, will use maximum resources to maintain a quality playing surface on a golf course.

For the homeowner and grounds manager, the newer turf-type tall fescue varieties offer a moderate quality level while requiring the use of fewer resources. Because quality level is vital to managers of sports turf, they will select turfgrasses that provide the best quality turf and use whatever resources are necessary to keep a quality turf.

Additional Factors

Additional factors to consider when selecting specific species and varieties of turfgrass include disease and insect resistance; cold, drought, and heat tolerance; salt tolerance; and color (Tables 5-1 and 5-2).

When selecting turfgrasses, the turf manager must decide between using a *monostand* (a single cultivar of one species) or a **polystand** (either a **blend** of two or more cultivars of one species or a **mixture** consisting of two or more species that may be blends).

If the grasses are compatible, blends and mixtures provide better overall defense against potential pest and environmental problems than does a monostand of a single cultivar. If disease attacks a monostand, the entire turf often will be damaged. In a polystand consisting of resistant cultivars or species, the resistant plants will fill in the spaces left by plants lost to disease or other problems.

SITE PREPARATION

A major goal in establishing a turf is site preparation, in particular, soil preparation. The more time spent in preparing the planting bed, the fewer problems in the future. Too often the turf manager's budget is reduced in the area of soil preparation. Unfortunately, the turf manager or customer pays for this budget cut later with increased maintenance practices, such as core aerification to loosen a tight clay soil.

Soil Preparation

Soil preparation involves:

- control of weeds and old grasses
- debris removal
- grading and shaping
- irrigation and drainage installation
- optional soil amendments
- final planting bed preparation

Control of Weeds and Old Grasses. To avoid problems later, the turf manager should attempt to control weeds and undesirable grasses prior to performing any other preparation activities. The goal is not to eliminate *all* weeds and weed seed from a site; it is to attempt to suppress weeds long enough for the new turf to grow and be capable of competing against any weed growth.

Weed control traditionally involved using a combination of selective herbicides and mechanical control over a long period of time. On some special turf areas, such as golf greens, turf managers used soil fumigation to control not only weeds but other potential turf pests. Current practice involves the use of glyphosate (Monsanto Roundup®, Ortho Kleenup®) and other non-selective herbicides that have a low soil residual.

Table 5-1 Characteristics of Common Cool-Season Turfgrasses

	Creeping Bentgrass *Agrostis palustris*	Kentucky Bluegrass *Poa pratensis*	Perennial Ryegrass *Lolium perenne*	Fine Fescue *Festuca rubra* ssp	Tall Fescue *Festuca arundinacea*
Establishment rate	Medium	Slow to medium	Fast	Medium-fast	Medium-fast
Recuperative ability	Best	Good	Poor to fair	Poor to fair	Poor to fair
Wear resistance	Poor	Fair	Fair to good	Fair	Good
Cold tolerance	Excellent	Excellent	Poor to fair	Fair	Fair
Drought tolerance	Poor	Good	Fair	Very good	Very good
Shade tolerance	Fair to good	Poor	Poor	Excellent	Fair to good
Salt tolerance	Best	Poor	Fair	Poor	Good
Maintenance level	High	Medium to medium-high	Medium	Low to medium	Low to medium
Fertility requirement	Highest	Medium to medium-high	Medium	Low	Medium
Disease problems	High	Medium	Medium	Medium	Low
Thatch tendency	High	Medium	Medium	Medium	Low

Table 5-2 Characteristics of Common Warm-Season Turfgrasses

	Bermudagrass *Cynodon dactylon*	St. Augustine *Stenotaphrum secundatum*	Buffalograss *Buchloe dactyloides*	Centipedegrass *Eremochloa ophiuroides*	Zoysiagrass *Zoysia* sp.
Establishment rate	Very fast	Fast	Moderate	Moderate	Very slow
Recuperative ability	Excellent	Good	Fair	Poor	Poor due to slow growth
Wear resistance	Very good to Excellent	Fair	Fair	Poor	Excellent
Cold tolerance	Poor to fair	Least tolerant	Fair	Very poor	Fair
Heat tolerance	Excellent	Excellent	Excellent	Excellent	Excellent
Drought tolerance	Excellent	Fair	Excellent	Poor	Excellent
Shade tolerance	Very poor	Excellent	Poor	Fair to good	Good
Salt tolerance	Good	Good	Good	Poor	Good
Maintenance Level	Medium to high	Medium	Low	Low	Medium
Fertility requirement	High	Medium	Low	Low	Medium
Disease problems	High	High	Low	Low	Medium
Thatch tendency	High	High	Medium	Medium	Medium to high

Glyphosate usually will control almost all plants, both weeds and original grasses, in one to three applications over a two-week (one application) to six-week (three applications) period. Ten to fourteen days following the last application of glyphosate, the turf manager can continue establishment. If the soil is of high quality and requires little additional work, the turf manager can seed directly through the dying vegetation.

Debris Removal. Debris left in the soil can lead to many problems later. The turf manager should attempt to remove all debris from the soil, including rocks, tree stumps, and construction materials.

Almost all debris can disrupt soil water movement, resulting in localized dry spots and poor growth. Rocks may rise to the surface from frost heave and result in costly mower damage, excavation of the rock, and repair of the turf.

Leftover tree stumps and construction wood can lead to problems of fairy ring, a surface complex of **fungus** organisms responsible for organic matter decomposition. The organisms tie up water and nutrients while they decompose organic matter. This tie-up results in rings of weak or dead turf where the fungus is most active. Once established, fairy ring is very hard to control.

Grading and Shaping. After debris removal, the turf manager can develop a rough-graded surface to provide the proper slope and contours for adequate surface-water drainage. Around buildings and other structures, the contour should slope away to eliminate potential water damage. On sports turf, the contour should slope to carry water away from any high traffic areas such as home plate on a baseball diamond.

Soil Amendments. Based on soil test results or other specifications, the turf manager should next incorporate any soil amendments such as **lime**, fertilizers, organic matter, or water-saving polymers. The turf manager must take care during mixing so as not to destroy soil structure through excessive tillage.

Irrigation and Drainage Installation. Too often, irrigation systems and subsurface drainage are afterthoughts to turf establishment. The homeowner or corporate client often does not realize the need for these systems until the turf is either dying from lack of available water or covered with standing water.

By installing these systems before establishment, surface damage caused by trenching is easy to correct. A pre-installed irrigation system provides the water supply necessary for growth of new turf.

Final Plant Bed-Preparation. Just before establishment, the turf manager's crew prepares the final planting surface. This involves grading the soil to form the final contours and cultivating to develop a soil structure that will be conducive to turfgrass growth.

To avoid destruction of any soil structure, the turf manager should avoid excessive grading and tillage using power equipment. Overworking the soil can result in crusting and surface compaction that may be detrimental to turfgrass root growth. Hand raking and shaping may be necessary when developing high-quality turf areas.

Site preparation is the same for any method of planting. To avoid excessive grade levels, however, the turf manager must account for the extra soil attached to sod or **plugs**. This involves lowering the surface grade, particularly near any hardscapes such as sidewalks and driveways, to accommodate the extra soil brought in with the turf.

ESTABLISHMENT BY SEED

Most cool-season turfgrasses and some warm-season turfgrasses can be established by seed. Several factors come into play when preparing to establish a turf by seed. These factors primarily involve proper environmental conditions that are favorable to both seed germination and young grass plant growth.

Seed Establishment Factors

Factors in establishing a turf by seed are:

- favorable soil temperature and moisture for germination
- favorable growing conditions until plants start to mature
- anticipated pest problems
- any additional germination requirements

Favorable Soil Temperature and Moisture for Germination. For any seed to germinate, several conditions must be met. Temperature and moisture are two critical factors in the germination process.

The soil must be warm enough to activate various physiological processes in the seed. A soil that is too cold or two warm will slow the processes down and can prevent them from occurring at all.

There must be adequate moisture in the soil to start and continue the germination process. For many types of seed, not just turfgrass, water is the trigger to start the germination process. It is important to realize that inadequate moisture levels may be sufficient to start germination only to kill the seeding later due to a shortage of water.

Favorable Growing Conditions until Plants Start to Mature. Even though favorable conditions may exist at the time of germination, unfavorable weather may occur before the turfgrass plant can reach a level of tolerance. This will result in severe setback or death of a turf.

As an example, a Kentucky bluegrass stand may fail from planting too late in the fall because the young plants cannot tolerate the freezing conditions. It is critical that seeds have adequate growing conditions until maturity.

Anticipated Pest Problems. Many turfgrass pests are more active at certain times of the growing season. For example, weed seed germination is greater at some periods of the season than at others. To avoid competition between turf and weed (wherein the weed often succeeds), the turf manager should plant seed when the incidence for weed germination is low.

Many insects are also far more active at certain times of the year than at others. As with weeds, the turf manager should plant seed when the chances for damaging insect problems are low.

Because many disease organisms are always present in the soil, the turf manager must be alert for possible disease attack on young turfgrass seedlings. Proper cultural practices, particularly watering, helps to reduce the incidence of disease. If necessary, the turf manager may apply **fungicides** as a preventive or curative measure.

Any Additional Germination Requirements. Some species of turfgrass require additional conditions to break seed **dormancy** and initiate germination. Light and cold temperatures are two factors sometimes needed to trigger germination. For example, to improve germination of buffalograss (*Buchole datyloides*) seeds, the hard hull surrounding the seed requires additional processing by dehulling and chilling.

General Guidelines for Best Times to Plant

The best planting time for cool-season grasses is generally late summer through early fall. At this time, the soil has warmed all summer and the rains and cooler, optimum growing temperatures of fall are just ahead. Also, most weeds have completed their growth, and many insects are going into a less active stage. Cool-season grass seeds can also be planted in early spring; but there is a greater risk of weed, insect, and disease problems along with the potential for rapid onset of hot, dry weather.

The best time to plant warm-season grasses is the onset of early summer. At this time, temperatures are entering an optimum growth range for these grasses. Also, the seeds are usually vigorous enough to out compete many weeds. It is best to avoid late summer through early fall seeding of warm-season grasses because of the increasing conditions in the fall that trigger dormancy. Many cool-season weeds are also germinating at this time.

VEGETATIVE ESTABLISHMENT

There are four methods of vegetative turf establishment. For those familiar with other landscape plants, vegetative establishment is a means of transplanting grasses that is similar to the transplanting of trees, shrubs, and vines.

Sodding

Sodding involves cutting and replanting large strips of turf. The process is very much like laying carpet or floor tile. Sod is often sold in strips of one square yard (0.836 m^2) usually measuring 4.5 feet (1.37 m) by 2 feet (0.61 m). Other sizes are available, including rolled sod strips that are several feet long by 2 feet (0.61 m) wide. Only a small amount of soil generally remains with the turfgrass in sod. For establishment on special root zones, such as all sand sports fields, some sod farms will wash the sod to remove all soil from the roots. This prevents soil-texture-interface problems that may disrupt the use of special root zones such as a Prescription Athletic Turf® (PAT) system.

Sodding enables a turf manager to establish a uniform, mature turf stand in a matter of weeks instead of the months required in establishment by seed. Unlike seeding, a turf manager can establish sod at any time conditions are favorable for the growth of the turfgrass.

The disadvantages is the high cost of sod. When the total cost of establishment (including costs for post-establishment care required to obtain a mature turf) is considered, however, sodding may be cost effective for areas up to one acre (0.41 hectare), or the size of a football field.

Plugging

Plugs are blocks or cylinders of turf with a large amount of soil attached. The depth of a plug is often the same size as the diameter or side length of the plug. A turf manager will use plugging to establish slow-growing turfgrasses such as Zoysia or to repair damaged turf areas such as golf greens.

Plugs are usually 2 inches (5.08 cm) in diameter or square by 1.5 inches (3.8 cm) to 2 inches (5.08 cm) deep. For establishment purposes, the turfgrass is often sold as sod strips. The plugs are cut from the strips and planted on a 4 inch (10.16 cm) to 6 inch (15.24 cm) spacing.

When repairing turf areas such as damaged golf greens, the size of the plug may be much larger in diameter. Turf managers can use special plugging tools to remove damaged turf and replace it with healthy turf. The tool allows for exact fit of the new turf plug into the existing turf to avoid creating an uneven playing surface.

Stolonizing and Sprigging

In any turfgrass with stolons or rhizomes, buds at the nodes of these lateral stems can lead to new plants. By cutting the stems in the internodes between the nodes, these stem pieces can be replanted to vegetatively establish a turf.

Stolonizing and sprigging are similar practices involving the planting of chopped pieces of stolons or rhizomes in a manner similar to seeding. Both methods work for turfgrasses that do not produce viable seed such as the hybrid bermudagrass.

Stolonizing involves scattering the stem pieces over the soil surface, commonly called broadcasting, partially covering the pieces with soil, and rolling for soil contact. Sprigging involves planting the stem pieces in holes or furrows. Stolonizing allows for quicker establishment, though there is a greater chance for drying.

CALCULATING MATERIALS NEEDS

For the turf manager, a successful establishment activity will result in a mature, uniform turf in the shortest time possible. It is important for the turf manager to determine the amount of materials needed to avoid difficulty in establishing the turf. Miscalculation of the amount of seed or sprigs needed could result in a thin turf from too few plants. If the seeding rate is too high, loss could occur from a disease attack on plants stressed from overcrowding. Too little sod will result in gapping holes in a customer's lawn. Too much sod will be a waste of materials because sod is live plant material that cannot be stored like seed.

Determining Seed Amounts

Even though a turf can be established using very small amounts of seed, the recommended seeding rates match an anticipated mature turf stand of ten to twenty-five plants per square inch. The seeding rates of major turfgrasses are lised in Table 5-3.

TABLE 5-3 Seeding Rates of Major Turfgrass Species Including Seed Numbers

Turfgrass Species	Seeding Rate in Pounds per 1,000 sq. ft.	Approximate Number of Seeds per lb.	Approximate Number of Seeds per sq. in. Based on Rate and Seed Number
Creeping Bentgrass	0.5–1.5	5,000,000–7,000,000	17–23
Kentucky Bluegrass	1–2	1,000,000–2,200,000	7–31
Fine Fescue	3–5	350,000–600,000	7–21
Tall Fescue	5–9	170,000–220,000	6–14
Perennial Ryegrass	5–9	220,000–250,000	8–16
Bermudagrass	1–2	2,000,000–2,500,000	14–35
Zoysiagrass	1–3	1,000,000–1,300,000	7–27
Centipedegrass	0.25–2	800,000–900,000	1–13
Buffalograss	3–7	40,000–100,000	1–5

The rates will vary depending on the:

- growth habit of the grass
- environmental conditions at planting time
- rate of fill-in desired
- seed availability and cost of seed
- desired distribution of grasses in a turf

Growth Habit of the Grass. Creeping grasses, with their lateral rhizome or stolon growth, fill in much faster than bunch-type grasses, which depend on tiller growth only. Lower seeding rates can be used with the creeping grasses that exhibit rapid **lateral growth**.

Environmental Conditions at Planting Time. Optimum environmental conditions result in high germination percentages. As conditions become less desirable, germination rates decrease. The turf manager can increase seeding rates to compensate for lower germination percentages.

Rate of Fill-in Desired. The higher the seeding rate, the quicker the fill-in of the turf due to the large number of emerging seedlings. Because of the stresses that may be created by overcrowding, however, there is an increased risk of loss from disease attack.

Seed Availability and Cost of Seed. As with any agricultural seed supply, several conditions influence the availability of a turfgrass seed. The seed may be difficult to produce, which limits its availability. Centipedegrass and buffalograss, for example, are two turfgrass species that have fairly low seeding rates due primarily to lack of availability. With new varieties of turfgrass, there may be limited availability until the producers can grow ample amounts of seed. Weather can also limit the seed producer's ability to produce ample supplies.

Cost of seed is often connected to availability. It is a classic situation of supply and demand. When there is limited supply and high demand, a seed producer or distributor can charge higher costs. Another factor that may result in higher cost is the expense incurred by the seed producer in

TABLE 5-4 Percent Species by Weight Versus Percent Species by Seeds per Pound

Species	Percent Species by Weight	Seeds per lb.	Percent Species Based on Seeds per lb.
Kentucky Bluegrass	50%	1,600,000	86.49%
Perennial Ryegrass	50%	250,000	13.51%

growing, harvesting, and processing seed. Some varieties require complicated breeding techniques to maintain the desired variety characteristics.

Although doubling the seeding rate will double the cost of seed, this will not significantly increase the cost of the establishment project. Depending on the species, the cost of seed will likely be a small percentage of the total operation costs.

Desired Distribution of Grasses in a Turf. For monostands or blends of one species, the seeding rates given in Table 5-3 will be appropriate. When dealing with mixtures, however, the number of seeds per pound plays a very critical factor in the final distribution of different species in the turf. Mixing together one pound of Kentucky bluegrass with one pound of perennial ryegrass, for example, will result in a distribution of far more Kentucky bluegrass than perennial ryegrass, as shown in the table comparing seed weight to seed count (Table 5-4).

Determining Sod

When sodding, complete coverage is necessary to obtain a uniform turf. Unlike other methods of establishment, there is no adjustable rate for sodding.

Because sod is usually sold on a square-yard basis, determining how much sod is needed is just a matter of converting the area to be sodded into square yards. The resulting figure is the amount of sod the turf manager will need to buy from a sod farm.

When buying sod, the turf manager should attempt to buy sod grown on a soil that is compatible with the soil underlying the area to be sodded. A wide difference between this soil and the soil clinging to the sod may lead to rooting problems and water movement problems.

Determining Sprig Amounts for Sprigging and Stolonizing

Determining the amount of **sprigs** required for sprigging or stolons required for stolonizing is very similar to determining the amount of seed required for a new turf area. Sprigging usually requires one to four bushels per 1,000 square feet (1.25 cubic feet per bushel) and stolonizing requires five to ten bushels per 1,000 square feet. As with seeding, the higher rates provide quicker establishment but also increase costs.

POST-PLANTING CARE

Water is probably the most critical factor in establishing a new turf. Seeds, sprigs, and sod all have limited root systems during the initial period of growth. Light, daily watering often is required to prevent drying out of the plants. As the plants mature, the watering schedule can be changed to a more routine irrigation schedule.

For seeding areas and some sprigged/stolonized areas, a light mulch will reduce drying problems and stabilize the soil to prevent erosion. Several materials are available for use as a mulch, with straw being the most popular. The average rate of straw needed for seeded areas is two bales (approximately 80 pounds) per 1,000 square feet (39 kg per 100m^2).

ACTIVITIES

Establishing a New Turf Area

Establish a new turf stand on a designated area. Perform two or more of the different methods of establishment: seeding, sodding, plugging, sprigging, and stolonizing.

Make records as to date of planting, germination or growth, and time to reach a mature stand. Keep notes as to the different methods of establishment as well as any post-planting care activities. Use the establishment activities record (Figure 5-1) for a guide.

Seed Mixture Calculations

Determine the amount of seed needed to obtain the following 1-pound-per-1,000-square-foot mixtures based on percentage species factors (as opposed to weight factors).

- a sunny/shady mixture of 50% Kentucky bluegrasss and 50% fine fescues
- a sports-turf mixture of 65% Kentucky bluegrass and 35% perennial ryegrass
- a general contractor's mixture of 50% Kentucky bluegrass, 30% fine fescues, and 20% perennial ryegrass
- A "tough turf" mixture of 80% dwarf turf-type tall fescue and 20% Kentucky bluegrass
- a shade mixture of 60% fine fescues and 40% shade-tolerant Kentucky bluegrass
- a winter overseeding mixture of 50% fine fescue and 50% perennial ryegrass

Using a Computer. Develop a spreadsheet to calculate the various mixtures.

On-Line Companion. U05A01.WKS—Seed mixture spreadsheet

```
West Administration Entryway Lawn

Seed: Campus Green (50% Kentucky Bluegrass,
25% Perennial Ryegrass, 25% Fine Fescue)
Rate: 37 lb. @ 3 lb/1,000 sq. ft.
Date Planted: August 15, 1998

Germination: PR & FF @8/26/98, KB @9/5/98

First Mowing: 9/12/1998

Problems with spotty annual bluegrass some areas.
A few broadleaf weeds controlled with post-emerge
herbicide after fourth mowing.
Mature: May, 1999
```

Figure 5-1 Establishment activities record

	Test 1	Test 2	Test 3	Test 4	Test 5
Seed Mix (Species and Cultivars)					
Spreader Type (Rotary/Drop)					
Spreader Brand and Manufacturer					
Suggested Spreader Setting					
Average Seeds per sq. in.					

FIGURE 5-2 Seed application uniformity table

Spreader Application Uniformity

Equipment.

- turfgrass seed (with labeled seeding rates)
- drop, gravity, rotary, or centrifugal spreader
- square-inch gauge consisting of an index card with a one-square-inch cut-out

Procedure.

1. Set the spreader to apply the recommended rate as given on the seed tag or in the spreader manufacturer's operation manual.
2. Apply the seed over a small area.
3. Using the square-inch gauge, count the number of applied seeds in a square inch. Sample several areas. With mixes, identify species during counting.
4. Average the samples to obtain a single average value.
5. Fill in the seed application uniformity table (Figure 5-2) with your data.

Does the recommended spreader setting apply the seed at a rate to match the suggested number of seeds per square inch?

EXERCISES

Turfgrass Species and Cultivar File

Create a reference file of different cultivars of various turfgrass species. Note such characteristics as color, pest resistance, texture, fertility requirements, and general maintenance requirements. For information resources, use the results from the National Turfgrass Evaluation Program and turfgrass supplier's promotional literature.

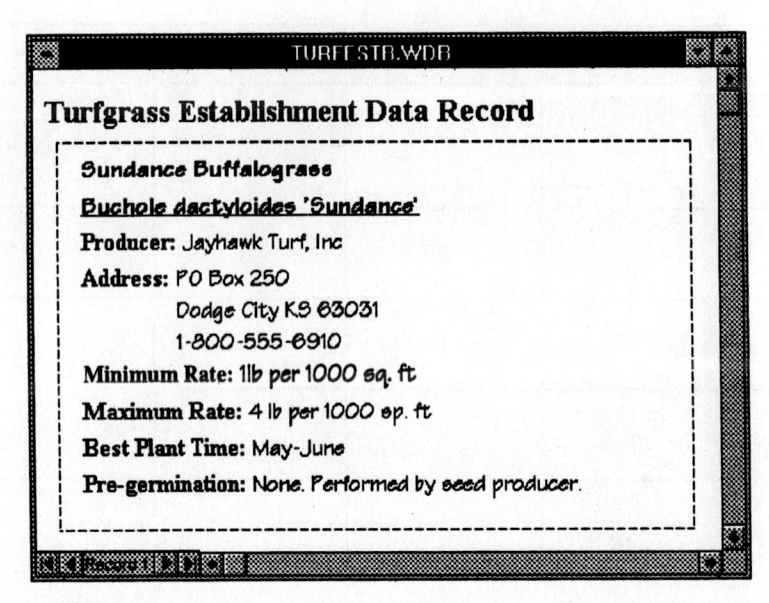

```
                    TURFESTB.WDB
Turfgrass Establishment Data Record
┌─────────────────────────────────────────────┐
│ Sundance Buffalograss                       │
│ Buchole dactyloides 'Sundance'              │
│ Producer: Jayhawk Turf, Inc                 │
│ Address: PO Box 250                         │
│          Dodge City KS 63031                │
│          1-800-555-6910                     │
│ Minimum Rate: 1lb per 1000 sq. ft           │
│ Maximum Rate: 4 lb per 1000 sp. ft          │
│ Best Plant Time: May-June                   │
│ Pre-germination: None. Performed by seed producer. │
└─────────────────────────────────────────────┘
```

FIGURE 5-3 Sample turfgrass species/cultivar record

Using Paper. Develop a series of index cards showing the different turfgrass species and cultivars. Record one species/cultivar per card.

Using a Computer. Develop a database of the different turfgrass species and cultivars (Figure 5-3). Use the query, searching, and sorting features to find species and cultivars to meet the needs of a particular turf site.

On-Line Companion. U05E01.WDB—Turfgrass species/cultivar selection database

Establishment Calculations

Develop a means of calculating the amount of materials needed to establish a turf area. Extend the calculations to develop a "bid for installing a lawn" using the various methods of establishment (seed, sod, sprigs, plugs). Use the area calculations from previous exercises.

Using Paper. Develop a paper form that will take a technician or customer through the steps necessary to determine the amount of materials needed. Complete a form for each of several different turf areas.

Using a Computer. Develop a spreadsheet of calculations including a lookup table of the different seeding rates as part of the calculations (Figure 5-4). As an alternative, generate a report combining the information from various data files in a relational database setup. If the various software packages support integration, merge the spreadsheet calculated bid into a wordprocessor document as part of a proposal letter to a customer.

On-Line Companion. U05E02.WKS—Establishment rate spreadsheet

	A	B	C	D	E	F	G
1							
2	Area (sq. ft)	37,000		Grass	Min	Max	Cost/lb
3	Grass	Kentucky Bluegrass		Bermudagrass	1	2	$0.25
4	Min Total Amount (lbs)	37		Buffalograss	3	7	$0.75
5	Max Total Amount (lbs)	74		Centipedegrass	0.3	2	$0.85
6	Min Total Cost	$5.55		Creeping Bentgrass	0.5	1.5	$0.50
7	Max Total Cost	$11.10		Fine Fescue	3	5	$0.08
8				Kentucky Bluegrass	1	2	$0.15
9				Perennial Ryegrass	5	9	$0.03
10				Tall Fescue	5	9	$0.07
11				Zoysiagrass	1	3	$0.65

FIGURE 5-4 Sample establishment calculations form

UNIT 6

Fertilization

Turf fertilization is the practice of applying elemental nutrients as fertilizers to improve turfgrass growth. The turf manager often needs to fertilize because of insufficient elemental nutrients in the soil. Even when there are adequate amounts of nutrients for moderate turfgrass growth, the turf manager still may apply a fertilizer to enhance growth for a particular purpose. A good example is a sports field where applications of fertilizer may be used to ensure that the turf is ready for play at the appropriate time.

Fertilization, or the lack of it, has a major impact on other turf management practices. No fertilization or a limited fertilization program can result in a weedy turf. At the extreme, excessive fertilizer applications can encourage several diseases as well as increase mowing and irrigation requirements. Use of fast-release fertilizers can lead to thatch problems that may require extensive cultivation and mechanical removal. A slow-release material may not provide enough nutrient to meet plant needs.

A well-balanced fertilizer program can result in a strong, healthy turf that is capable of competing against weeds while being tolerant of environmental extremes and pest problems. A strong turf also can withstand mowing, itself a stress on the turf plant, and better utilize water from irrigation.

DETERMINING FERTILIZER NEED

There are several means of determining fertilizer need. The element to be enhanced usually has an impact on determining need. Due to the ever-changing nature of nitrogen in the soil, turfgrass growth, recommended fertilizer rates, and possibly tissue tests are better means of determining need than are soil tests. For the other nutrients, primarily phosphorus and potassium, the turf manager can use various soil tests along with recommended rates as a ratio to nitrogen to determine fertilizer need.

Growth and Color

In comparison to woody ornamentals such as trees, turfgrasses often exhibit reduced growth rates quite quickly when nutrient deficiencies, particularly of nitrogen, start to occur. For golf course superintendents, a common indicator of nutrient deficiency is the number of baskets of clippings removed during mowing of golf greens. As the number decreases, it is a good indication to the superintendent that there is a need to apply more fertilizer to maintain growth.

Color is another common indicator of nutrient deficiencies (Table 6-1). A yellowing of leaves indicates a nitrogen or potassium deficiency; a purple discoloration is often an indication of phosphorus deficiency; while a bright-yellow color is a good indication of iron deficiency in the plant resulting from tie up of iron in high pH soils.

Changes in growth and color do not necessarily indicate a need for more fertilizer, however. Several other factors can contribute to discoloration and reduced growth, including time of season, pollution, misapplication of pesticides, and water-logged soils. If changes in growth or color occur the turf manager should quickly check for other possible problems and use other methods of determining fertilizer need before applying additional fertilizer.

TABLE 6-1 Nutrient Deficiency Symptoms

Element	Deficiency Symptoms
Nitrogen	Older leaves turn yellow, reduced shoot growth
Potassium	Interveinal yellowing especially on older leaves and leaf tips, margins scorched
Phosphorus	Older leaves dark green first, then appear purple or reddish
Calcium	Deficiency rare in turf
Magnesium	Interveinal chlorosis, a striped appearance, cherry red margins
Sulfur	Yellowing of older leaves
Iron	Bright interveinal yellowing of new leaves
Manganese	Rare in turf
Copper	Never a problem
Zinc	Rare in turf
Boron	Rare in turf
Molybdenum	Rare in turf
Chlorine	Never a problem

Interpreting Soil Test Results

Soil testing is a common means of determining fertilizer need for most mineral nutrients except nitrogen. Most soil tests give the amount of available element in the soil along with a recommended rate of application to meet the needs of the turfgrass plant. Some soil test results give two rates: one to bring nutrient rates to an acceptable level and one for "maintenance," to meet the needs of the turfgrass on a regular basis (Figure 6-1).

Soil tests are also valuable in that they usually include soil reaction or pH. In many turf situations, correcting the pH will release nutrients that have been unavailable to the plant without the need for additional fertilizers.

Recommended Fertilizer Rates

Different turfgrasses have different nutrient needs. Kentucky bluegrass often requires two to four times as much nitrogen as needed by fine fescues. Furthermore, cultural practices, use of the turf, and environmental conditions all have an impact on the amount of nutrients needed by the turf.

The changing nature of nitrogen makes it difficult for the turf manager to use the results of soil tests in determining nitrogen need. The nitrogen in a soil often changes form or is lost before the turf manager receives the soil test results from a soils testing lab. Turf specialists usually base nitrogen fertilizer rates on the amount of nitrogen needed for adequate growth of turfgrass within a species (Table 6-2). Turf specialists establish these rates from on-going research by continually studying turfgrass growth with various rates of fertilizers.

Turf specialists often give the rate of phosphorus and potassium required as a ratio related to the nitrogen fertilizer rate. This is known as the **fertilizer ratio**. The ratio usually ranges from 6-1-1 (6 parts nitrogen, 1 part phosphorus, and 1 part potassium) to 3-1-2. Several factors determine the ratio used by the turf manager, including soil conditions, time of year, and culture. Using soil test results for phosphorus and potassium, the turf manager can also adjust the ratio to compensate for the existing level of phosphorus and potassium in the soil.

60 Unit 6

**TURF AND ORNAMENTAL
SOIL TEST AND RECOMMENDATION REPORT**

REPORT TO: T997
A TURF CONSULTANT
325 MAIN STREET
ANYTOWN, USA 44999

SUBMITTED BY/FOR: A STATE COLLEGE
PHYSICAL PLANT DEPT
COLUMBUS, OH 43999

CLC LABS
325 VENTURE DRIVE
WESTERVILLE, OHIO 43081
614 888-1663

08/12/99 C
1000.055

RESULTS OF ANALYSIS / CALCULATED VALUES

REPORT REF. NUMBER	LAB. NO.	Soil pH	Buffer pH	P	K	Ca	Mg	Cation Exchange Capacity	K	Ca	Mg	H	Na	Fe	Mn	Zn	Cu
1	137300	8.0		26	150	4651	338	13.2	1.5	88	11	0.0		85	3	3.0	2.5
2	137301	7.8		22	138	4774	355	13.6	1.3	88	11	0.0		101	2	3.2	2.7
3																	
4																	
5																	
6																	
7																	
8																	
9																	
10																	
11 AVERAGE RESULTS →				24	144	4713	347	13.4	1.4	88	11	0.0		93	3	3.1	2.6

DISPLAY OF AVERAGE RESULTS

SURPLUS						*			*								
HIGH					*	*		*	*	*				*			*
MEDIUM				*	*	*	*	*	*	*				*		*	*
LOW			*	*	*	*	*								*	*	*

SAMPLE INFORMATION / FERTILIZER RECOMMENDATIONS IN LBS. PER 1,000 SQ. FT.

REPORT REF. NUMBER	SAMPLE IDENTIFICATION	PLANT TYPE	AREA TYPE	FERT/MAINT. LEVEL	LIME LBS/M	LIME TYPE	NITROGEN	APP. TYPE	P₂O₅	K₂O	Mg	Fe	Mn	Zn	COMMENTS
1	STADIUM SOUTH END	KY.BLUE/PER.RYE	FOOTBALL FLD	HIGH			4.0 -5.0	S	4.0	2.0	1.0		0.04		See All
2	STADIUM NORTH END	KY.BLUE/PER.RYE	FOOTBALL FLD	HIGH			4.0 -5.0	S	4.0	2.0	1.0		0.04		See All
3-10															
11 RECOMMENDATIONS FOR AVERAGE RESULTS →							4.0 -5.0	S	4.0	2.0	1.0		0.04		See All

SEE COMMENTS ON REVERSE SIDE

DUE TO VARIATIONS IN WEATHER, SOIL CONDITIONS AND CULTURAL PRACTICES, NO WARRANTY EITHER EXPRESSED OR IMPLIED IS MADE WITH RESPECT TO PLANT PERFORMANCE.

FIGURE 6-1 Sample soil test results

Fertilization

UNDERSTANDING YOUR SOIL TEST REPORT

ANALYTICAL RESULTS

A. **SOIL pH**: a measure of acidity or alkalinity in water according to the following: pH 7.0 is neutral, pH <7.0 is acidic and pH>7.0 is alkaline. Most turf and ornamentals prefer a pH of 6.5 - 7.5. Certain acid-loving shrubs prefer a pH<6.0.

B. **BUFFER pH**: a measure of the soil's ability to acidify a buffered solution. Buffer pH is used to determine the lime requirement. Water pH alone should not be used to determine lime needs.

C. **PHOSPHORUS - P**: shows the amount of available phosphorus (Bray 1) found in the sample expressed in pounds of available phosphorus per acre.

D. **POTASSIUM - K**: shows the amount of available potassium (Exchangeable-K) found in the sample expressed in pounds of potassium per acre.

E. **CALCIUM - Ca and Magnesium - Mg**: shows the amount of available (exchangeable) calcium and magnesium found in the soil expressed in pounds of calcium or magnesium per acre, when these tests are requested.

F. CALCULATED VALUES

1. **CATION EXCHANGE CAPACITY - CEC**: a calculated value showing the relative nutrient holding capacity of the soil for the cations K^+, Ca^{++}, Mg^{++}, H^+ (hydrogen), when the Complete Test Package is requested. CEC will include the Na^+ (sodium) cation, when a Sodium test is requested.

Typical CEC Ranges	Soil Texture	Relative Nutrient Holding Capacity	CEC
0-12	Coarse (sandy)	Low	<10
8-30	Medium (loamy)	Medium	10-22
22-40+	Fine (clayey)	High	>22
30-50+	Organic soil		

2. **% BASE SATURATION** - calculated values showing the percentage of the CEC occupied by each cation, when the Complete Test Package is requested. Most turfgrasses and ornamentals perform best when the cations are in balance in the ranges shown below.

Potassium - K 2-7%   When present: Hydrogen - H 0-5%
Calcium - Ca   65-85% When tested: Sodium - Na 0-5%
Magnesium - Mg 10-20%

G. **DISPLAY OF AVERAGES RESULTS** - the average results printed on line 11 are displayed graphically. This provides a visual interpretation of the relative status of the results of all samples shown.

H. **MICRO & SECONDARY NUTRIENTS** - the nutrient content of the soil is shown for each micro and secondary nutrient test requested expressed in pounds of available nutrient per acre. Relative values are shown below.

				lbs./acre		
RELATIVE VALUE	IRON (Fe)	MANGANESE (Mn)	ZINC (Zn)	COPPER (Cu)	SULFUR (S)	BORON (B)
LOW	<15	<10	<2	<0.5	<20	<0.5
MEDIUM	15-120	10-50	2-5	0.5-5.0	20-80	0.5-3.0
HIGH	>120	>50	>5	>5.0	>80	>3.0

I. **ORGANIC MATTER - OM**: shows the organic content of the soil (% by weight), when an organic matter test is requested. Most mineral soils contain 2-6% organic matter.
COMBUSTION TEMPERATURES:
OM2=360°C, OM3=440°C, OM4=750°C

J. **SOLUBLE SALTS - SS**: shows the soluble salts content of the soil when soluble salts test is requested. A soluble salts test is used to determine if excessive salts exist in the soil, typically from over-fertilization or de-icing compounds. The relative degree of plant injury is shown below.

Soluble Salts (mhos X 10^{-5})	Potential Plant Injury
0-75	Very Low
76-150	Low
151-200	Medium (sensitive plants may be injured)
201-300	High
+300	Very High (most plants injured)

K. NOTES:
1. Optimum levels of plant nutrients vary with plant type, use and fertility/maintenance level. These factors along, with soil test information are used to make specific fertilizer recommendations. Bar graphs show relative amounts for soils in general.
2. To convert lbs. of nutrient per acre to parts per million divide reported values by 2.
3. Results followed by a "+" are outside the normal test range. Actual values are higher than shown.

CLC LABS uses soil testing methods published by the NCR-13 Committee of the U.S.D.A. Agricultural Research Service. These methods are widely accepted and used by many state university soil testing laboratories. CLC LABS performs extensive in-house quality control procedures and maintains soil testing certifications by the U.S.D.A. ASCS in several states to insure accuracy of all results.

LIME AND FERTILIZER RECOMMENDATION COMMENTS

CAUTION! To avoid plant injury consult a professional in the turf or ornamentals industry or your County Cooperative Extension Service before using recommended fertilizers or lime. Split applications of recommended lime and/or nutrients may be appropriate to avoid injury.

1. **ALL RECOMMENDATIONS** represent a typical range for the plant type, its use and fertility/maintenance level as represented by the sample information provided and the soil test results. Actual fertility management, ie. rate and timing of application, nutrient source and application method, etc. may vary widely with different cultural practices.

2. **LIME RECOMMENDATIONS** are given in pounds (lbs.) per 1,000 sq. ft. or tons per acre of agricultural ground limestone (TNP > 90%). Recommendations are for the amount needed to correct acid soil conditions. Do not over apply lime to established turf areas. Incorporate recommended amount into the root zone at establishment.
LIME TYPE - When calcium and magnesium tests are performed, the lime type recommended will be indicated as high calcium lime (Ca) or high magnesium/dolomitic lime (Mg).

3. **NITROGEN RECOMMENDATIONS** are given in lbs. per 1,000 sq. ft. or lbs. per acre of actual nitrogen (N). APP TYPE - Recommendations for application frequency given on a per season (S) basis should be split into multiple applications. Recommendations may also be given on a per month (M) of growing season or month of establishment basis. When NEW/ESTB. is selected as the fertility/maintenance level then the establishment recommendations are for incorporation into the soil at the time of planting (preferred) or for application during the first month or more of establishment.

4. **PHOSPHATE RECOMMENDATIONS** are given in lbs. per 1,000 sq. ft. or lbs. per acre of P_2O_5. Recommendations are given as the annual requirement for maintenance; the corrective amount, if low; or the amount to be used during the establishment phase.

5. **POTASSIUM RECOMMENDATIONS** are given in lbs. per 1,000 sq. ft. or lbs. per acre of K_2O. Recommendations are given as the annual requirement for maintenance; the corrective amount if low; or the amount to be used during the establishment phase. Do not over apply potassium.

6. **OTHER NUTRIENT RECOMMENDATIONS** are given in lbs. per 1,000 sq. ft. per acre of elemental magnesium (Mg), iron (Fe), manganese (Mn) or zinc (Zn). Recommendations are given as the corrective amount for maintenance or the amount to be used during the establishment phase. Do not over apply micronutrients.

FIGURE 6-1 Sample soil test results *(continued)*

TABLE 6-2 Recommended Nitrogen Rates for Most Turfgrasses

Species	Pounds of Nitrogen per 1,000 sq. ft. per Growing Month
Creeping Bentgrass	0.5–1.3
Kentucky Bluegrass	0.4–1.0
Perennial Ryegrass	0.4–1.0
Fine Fescue	0.1–0.4
Tall Fescue	0.4–1.0
Bermudagrass	0.5–1.4
St. Augustine	0.5–1.0
Buffalograss	0.1–0.4
Centipedegrass	0.1–0.3
Zoysiagrass	0.5–0.8

When choosing a fertilizer rate, the turf manager will need to decide between the goals of *maintaining* grass and *growing* grass.

The goal of maintaining grass means applying just enough fertilizer to keep the grass healthy and of adequate visual quality. At the same time, the rates are low enough to keep growth to a minimum and to permit reduced mowing, irrigation, and other cultural practices.

The goal of growing grass means applying fertilizer at rates that will encourage the growth of grass that has the ability to quickly recover from wear and tear. This can result in a dense, higher-quality turf but will increase the need for mowing, irrigation, and other cultural practices, including disease control.

For home lawns, corporate grounds, college campuses, and utility turf, where the major purpose of the turf is aesthetic, the turf manager should aim toward using the lower recommended rates. The turf manager should use the higher recommended rates on high traffic turfs to maintain a quality turf surface. Sod farmers who produce crop for sale, should use the highest possible rates in order to obtain a marketable product in the shortest amount of growing time.

FERTILIZER MATERIALS

Fertilizer materials that supply nutrients are known as *carriers* or *sources*. These carriers consist of inorganic or organic chemical compounds that contain one or more of the nutrients. As an example, potassium nitrate (KNO_3) contains the major nutrients nitrogen and potassium. Milorganite®, a complex organic material from activated sewage sludge, contains the major nutrients nitrogen and phosphorus. As these compounds dissolve or decompose, the desired nutrients are made available to the plant as nitrates (NO_3), ammonia (NH_3), phosphates (PO_3), and potassium ions (K^+).

When a fertilizer consists of a combination of carriers to supply the three major nutrients of nitrogen, phosphorus, and potassium, it is called a **complete fertilizer**. The percentage composition of the three major nutrients in a fertilizer is known as the **fertilizer analysis**.

All states have laws requiring the fertilizer analysis to be prominently displayed on a fertilizer label as a *fertilizer grade*, that is, as Percent Nitrogen-Percent Phosphorus-Percent Potassium. For example, a common 10-10-10 garden fertilizer contains 10 percent nitrogen, 10 percent phosphorus, and 10 percent potassium. For some states, the grade is a minimum guaranteed analysis.

> **Slow Release Fertilizer with IBDU**
> **20-4-10**
> **50 LB**
>
> Total Nitrogen (N)..20%
> 9.0% Ammoniacal Nitrogen
> 5.0% Water-Insoluble Nitrogen
> 6.0% Water-Soluble Urea Nitrogen
> Available Phosphoric Acid (P2O5)............................4.0%
> Soluble Potash (K2O) 10.0%
> Sulfur (S)..13.5%
>
> Derived from ammonium sulfate, ammoniated phosphate, isobutylidene diurea, urea, and sulfate of potash. Potential acidity 1,300 lb. Calcium carbonate equivalent per ton.

FIGURE 6-2 Typical fertilizer label information

In addition, in some states, the fertilizer manufacturer must express the analysis of phosphorus and potassium as phosphorus pentoxide (P_2O_5) and soluble potash (K_2O). The turf manager must convert these expressed percentages to determine the elemental form. The percentage P_2O_5 is multiplied by 0.44 to obtain the percentage phosphorus; while the percentage K_2O is multiplied by 0.83 to obtain the percentage potassium.

Along with the prominent display of the fertilizer analysis, most fertilizer labels, particularly those on turf fertilizers, contain a more detailed breakdown of the various nutrients and sources in the fertilizer (Figure 6-2). Percentage of total nitrogen often is listed first followed by a breakdown of percentages, based on solubility, of the different forms of nitrogen.

SELECTING FERTILIZER MATERIALS

The turf manager must take several factors into account when selecting a fertilizer material. Of main concern is the release of nitrogen from fertilizer carriers for turfgrass growth. The turf manager must choose between a fertilizer with highly water-soluble nitrogen sources or slow-release nitrogen sources.

Generally, water-soluble nitrogen sources, or quick-release fertilizers, provide a fair amount of readily available nitrogen to the turfgrass plant. This allows for quick green-up and rapid growth. The turf manager must, however, use care in the application of water-soluble nitrogen sources, such as ammonium nitrate or urea. Application of large amounts of water-soluble sources result in a rapid flush of growth followed by a similarly rapid drop-off of growth. It is best to apply small amounts of water-soluble fertilizers on a consistent weekly or bi-weekly basis. In addition, water-soluble fertilizers are often salt-like compounds. Excessive application or spillage can result in *fertilizer burn*, which is death of the grass plant due to the **desiccating** effects of high-solute concentrations.

Slow-release fertilizers are compounds that break down slowly based on water availability, temperature, or bacteria decomposition. This slow breakdown provides low amounts of nutrients to the plant continuously over a fairly long period of time. This allows the turf manager to make only a few applications per season. The rate of release may not be fast enough, however, to meet the needs of the plant. A common problem with slow-release products occurs in the spring when the turfgrass is actively growing but conditions are not yet favorable for the release of nitrogen from these products. To compensate for the problem, manufacturers of turf fertilizer combine various amounts and types of water-soluble and slow-release products. The quick-release portion of the resulting product meets the growth demands of the plant in spring. As the quick-release portion of the product dissipates, the slow-release portion finally dissolves to a level in soil solution that is adequate to supply the plant on a long-term basis.

In addition to release rate, the turf manager must consider several other factors when choosing a fertilizer. One of these is the impact of fertilizer analysis on storage and handling. If using a fertilizer with a low analysis number (i.e., low percentage of nutrients) the turf manager will need more material than if using a product with a high analysis. For example, to apply one pound of nitrogen to a turf will require seven times as much Milorganite (6-2-0) as urea (45-0-0). For the turf manager, more material means the need for a larger area to store the material, as well as increased labor and equipment to handle the material.

Another factor the turf manager must consider is whether to choose a mixed fertilizer or a blended fertilizer. A mixed fertilizer contains different carriers combined to make a complete fertilizer. A good indicator of a mixed fertilizer is the presence of different-colored **granules**. In a blended fertilizer, the manufacturer pulverizes each source, then combines the material into uniform granules. Uniform-colored granules in a complete fertilizer often indicates a blended product. When applied using certain types of mechanical applicators, mixed fertilizers tend to segregate into bands of separate carriers because of differences in particle density. Because each granule in a blended fertilizer is a complete fertilizer carrier this "banding" does not occur. Blended fertilizers, therefore, provide a more uniform application of material. Most fertilizer manufacturers sell turf fertilizer as blended products for this reason.

TURF FERTILIZER APPLICATION

Accurate fertilizer application is important in the management of turf areas. Misapplication of fertilizer materials, either through over fertilization or under fertilization, can lead to many other problems including disease attack and weed encroachment. Accurate, efficient application of fertilizer is important to avoid going over a pre-established budget (especially given the current high costs of these products).

Recommended fertilizer rates are generally based on the recommended amount of actual nitrogen per unit area. In the United States, most turf fertilization rates are based on *pounds of element per 1,000 square feet*. Recommended phosphorus and potassium rates are based on a ratio to nitrogen and range from 2-1-1 to 4-1-2. Most turf fertilizers are already formulated to a desirable ratio.

Calculating Fertilizer Amounts

To properly apply a fertilizer material to meet recommended rates, the turf manager must make accurate calculations. These calculations involve converting the recommended nutrient rate into an equivalent fertilizer material rate. The amount of fertilizer to apply to a given area to meet a nutrient recommendation can be calculated using the formula:

$$\textbf{rate of fertilizer} = \text{rate of element} \times \frac{100}{\% \text{ element}}$$

where rate of element is usually *pounds of element per 1,000 square feet* or *kilograms per 100 square meters*.

To determine the amount of fertilizer to be applied over a given area:

$$\textbf{total amount of fertilizer} = \text{rate} \times \frac{\text{area}}{\text{unit area}}$$

where unit area is from the rate, usually 1,000 square feet.

When the total amount of fertilizer has been determined, appropriate cost calculations can be applied to determine the total cost of fertilizer.

Granular Fertilizer Calculation. Granular fertilizer calculations are straightforward, as rates are determined on a weight basis. To apply 0.75 pounds of nitrogen per 1,000 using an 18-5-9 granular fertilizer over a 23,000 square foot lawn:

1. Determine the rate of fertilizer.

$$\text{pounds of nitrogen per 1,000 square feet} \times \frac{100}{\text{percent of nitrogen in fertilizer}} =$$

pounds of fertilizer per 1,000 square feet

$$0.75 \times \frac{100}{18} = \textbf{4.17 lb. of fertilizer per 1,000 sq. ft.}$$

2. Determine the amount of fertilizer.

$$\text{pounds of fertilizer per 1,000 square feet} \times \frac{\text{total area square feet}}{\text{unit area square feet}} =$$

pounds of fertilizer over total area square feet

$$4.17 \text{ lb.} \times \frac{23,000}{1,000} = \textbf{95.91 lb. of fertilizer over 23,000 sq. ft.}$$

Liquid Fertilizer Calculation. For liquid fertilizers, an additional calculation must be applied because liquid fertilizer is measured in volume and fertilizer rates are given in weight values. Calculating the amount of liquid fertilizer requires knowing the weight of one volume unit (e.g., gallon, liter) of fertilizer. To apply 0.75 pounds of nitrogen per 1,000 square feet using an 18-5-9 liquid fertilizer over a 23,000 square foot lawn:

1. Find the weight of one gallon of fertilizer either by weighing one gallon or from information provided on the label. For this example, one gallon of liquid fertilizer weighs 12.5 pounds.
2. Determine the pounds of nitrogen in one gallon.

$$\text{pounds per gallon} \times \text{percent of nitrogen in fertilizer} = \textbf{pounds of nitrogen per gallon}$$

$$12.5 \times 0.18 = \textbf{2.25 pounds of nitrogen per gallon}$$

3. Determine the volume rate of fertilizer.

$$\frac{\text{pounds of nitrogen to be applied per 1,000 square feet}}{\text{pounds of nitrogen per gallon}} =$$

gallons of fertilizer per 1,000 square feet

$$\frac{0.75}{2.25} = \textbf{0.3 gallons per 1,000 sq. ft.}$$

4. Determine the volume amount of fertilizer.

$$\text{gallons of fertilizer per 1,000 square feet} \times \frac{\text{total area square feet}}{\text{unit area square feet}} =$$

gallons of fertilizer over total area square feet

$$0.3 \times \frac{23{,}000}{1{,}000} = \textbf{6.9 gallons of fertilizer over 23,000 sq. ft.}$$

Methods of Application

There are two major methods of applying fertilizer to a turf area. The most common method is dry application using granular fertilizers; liquid application is the next most common method. The method chosen by a turf manager depends on several factors including the size of the turf area, equipment, labor, and desired rate of turfgrass response.

Granular Applications. One can apply dry or granular fertilizer to a turf using a variety of spreading techniques. Water movement or mechanical action carries the material down to the soil surface. The fertilizer either dissolves into the soil solution for uptake by the roots, or is decomposed by soil organisms for later release for later release into soil solution.

Dry fertilizer application requires minimal equipment. Dry fertilizer also is easier to handle than liquid fertilizer. Through granular application of the right fertilizer carrier, the turf manager can achieve long-term release of nutrients while reducing applications and related application costs.

Liquid Applications. Because the fertilizer is already in water, liquid application is useful in some turf situations. With liquid application, there is no need to depend on rainfall, irrigation, and soil water to dissolve the fertilizer for uptake by the plant.

A major advantage of liquid application is the quick response from turfgrasses to the fertilizer. Liquid application is very useful in providing a turf with "a quick shot" until the slower granular fertilizer can start providing nutrients.

Fertigation. A variation of liquid application is **fertigation**, or *ferti*lizer-injected irri*gation*. During irrigation, a pump injects liquid fertilizer into the irrigation system, which results in a constant, low-level fertilization of the turf during irrigation. There are a number of drawbacks to fertigation. One of these is the extensive work required to properly design and install an irrigation system that will uniformly apply the fertilizer. There is also a risk of environmental and health hazards when using large scale fertigation. It is generally best to leave fertigation to nurseries and greenhouses.

ACTIVITIES

Comparison of Different Fertilizer Rates

On a level, open area of campus, set up plots for a comparison study of different fertilizers, including quick-release versus slow-release materials. Over the term or semester, apply standard quality ratings to evaluate the plots. If the time of year prevents outdoor study, perform the activity in a greenhouse with pots or flats of maintained turf.

Examination of Different Fertilizer Materials

Examine the various fertilizer samples provided by the instructor. Place samples of each material in small, self-locking bags for later reference. Label each bag with the name of the fertilizer product, manufacturer, and analysis.

Reading and Interpreting Fertilizer Bag Labels

Examine several different fertilizer labels and fill in the fertilizer products (Figure 6-3).

Visit with Fertilizer Company Representatives

With assistance from your instructor, visit with fertilizer company representatives to learn about their various fertilizer products and services.

Calculating Fertilizer Amounts and Costs

Solve the following fertilizer problems.

1. The turf manager wants to apply a 10-5-5 fertilizer at the rate of 1 pound of nitrogen per 1,000 square feet. How many pounds of fertilizer per 1,000 square feet should the manager apply to the turf?
2. If the area in question is a football field (120 yards by 33 yards), use the information from the previous problem to determine the total amount of fertilizer for the field.

	Fertilizer 1	Fertilizer 2	Fertilizer 3	Fertilizer 4	Fertilizer 5
Supplier					
Brand					
Analysis					
Net Weight					
Unit Cost					
% Nitrogen					
Nitrogen Source 1					
Nitrogen Source 2					
Nitrogen Source 3					
Nitrogen Source 4					
% Slow Release (WIN)					
% Fast Release (WS)					
% Phosphorus					
% P as P_2O_5					
% Potassium					
% K as K_2O					
Calcium Carbonate Equivalent					

FIGURE 6-3 Fertilizer products table

3. Each bag of 10-5-5 fertilizer weighs 60 pounds and costs $21.00. How much will it cost to fertilize the field in the previous problem?
4. The label on a 36-pound bag of 26-8-12 fertilizer lists coverage at 10,000 square feet. What is the rate of nitrogen per 1,000 square feet if you apply the fertilizer with a spreader at the recommended setting?

EXERCISES

Fertilizer Label File

Collect several different fertilizer labels. Include labels from fertilizers used to fertilize the campus turf areas. Photocopies of the labels are acceptable. Mount the labels on standard size paper if necessary.

Using Paper. Keep all labels in your three-ring binder for later reference. Develop a paper table similar to Figure 6-3 for recording and organizing fertilizer information.

Using a Computer. Keep all labels in your three-ring binder. If a scanner is available, scan the labels into a graphic file for later reference. Develop a spreadsheet table similar to Figure 6-3 for recording and organizing fertilizer information.

On-Line Companion. U06E01.WKS—Fertilizer information spreadsheet

Calculating Fertilizer Amounts

Using information from previous activities and exercises, calculate the amount of fertilizer needed to meet rates as assigned by your instructor for the various campus turf areas.

Using Paper. Develop a paper table to calculate the amounts of fertilizer needed for the different areas. Expand and enhance a copy of the tables you created in the Unit 2 exercise Site Inventory of Area Measurements to calculate the fertilizer amounts. Include totals with the calculations. The column-formatted three-ring binder spreadsheets used by accountants are a good source of preprinted tables. Keep the completed tables in your three-ring binder.

Using a Computer. Develop a spreadsheet table to calculate the amounts of fertilizer needed for the different areas (Figure 6-4). Expand and enhance a copy of the files created in the Unit 2 exercise Site Inventory of Area Measurements to calculate the fertilizer amounts. Total all the different amounts for each fertilizer.

On-Line Companion. U06E02.WKS—Fertilizer rates spreadsheet

Fertilizer Cost Analysis

Using the information from previous activities and exercises, determine the costs of the fertilizers determined in the exercise Calculating Fertilizer Amounts.

Using Paper. Enhance the paper tables to include total costs of the fertilizers.

Using a Computer. Modify the files from the previous exercise to calculate costs (Figure 6-5).

On-Line Companion. U06E03.WKS—Fertilizer cost analysis spreadsheet

```
                          FERTAMTS.WKS
    A │      B          │    C    │     D       │    E       │    F
  1
  2   │ %Nitrogen       │   32%
  3   │ Appl. Rate (lbs/1000 sq) │ 0.8
  4
  5   │ Campus Area     │ Lawns   │ Sports Field│ Work Areas │ Dorms
  6   │ Area (sq. ft)   │ 236,540 │  96,750     │  48,920    │ 78,350
  7   │ Applications per season │ 2 │ 3          │  1         │  2
  8   │ Total Fertilizer (lbs) │ 1182.7 │ 725.6 │ 122.3      │ 391.8
  9
 10   │                 │         │             │ Total Weight │ 2422
 11
 12
```

FIGURE 6-4 Sample fertilizer calculations table

```
                          FERTCOST.WKS
    A │      B          │    C    │     D       │    E       │    F
  1
  2   │ %Nitrogen       │   32%
  3   │ Appl. Rate (lbs/1000 sq) │ 0.8
  4   │ Bag Weight (lb) │   47
  5   │ Cost per bag    │ $19.95
  6
  7
  8
  9   │ Campus Area     │ Lawns   │ Sports Field│ Work Areas │ Dorms
 10   │ Area (sq. ft)   │ 236,540 │  96,750     │  48,920    │ 78,350
 11   │ Applications per season │ 2 │ 3          │  1         │  2
 12   │ Total Fertilizer (lbs) │ 1182.7 │ 725.6 │ 122.3      │ 391.8
 13   │ Bags needed     │  25.2   │  15.4       │  2.6       │  8.3
 14   │ Cost per season │  $502   │  $308       │  $52       │  $166
 15
 16   │                 │         │             │ Total Weight │ 2422
 17   │                 │         │             │ Total Bags   │ 52
 18   │                 │         │             │ Total Cost   │ $1,028
```

FIGURE 6-5 Sample fertilizer costs table

UNIT 7

Mowing

Mowing is the one practice that distinguishes a turf from other grass-covered areas. There are many grass-covered areas that receive fertilizers and irrigation water. Pasture land is a good example. It is, however, through regular, routine mowing that the turf manager achieves a smooth, uniform surface for aesthetic, functional, and recreational purposes.

MOWING PROGRAM

Of the three primary cultural practices, mowing often requires far more labor and equipment than does fertilization or irrigation. Mowing or, more importantly, improper mowing, can have an enormous impact on other turfgrass management practices. For example, setting a **cutting height** too low for a turf can lead to serious weed problems because of stress on the turfgrass and the ability of some weeds to tolerate close mowing.

Mowing Height

The key factor in developing a mowing program is the mowing or cutting height of a turf. Usually, mowing height refers to the **bench setting**, or the height of the mower blade(s) above a hard surface such as workbench. There is also the cutting height, or effective cutting height, or the height of the mower blade(s) above the soil surface, which will be a variable value due to turf conditions such as thatch thickness. Four interrelated factors influence mowing height:

- use of the turf
- turfgrass species and cultivar/variety
- time of the season
- health of the turfgrass

Use of the Turf. In most turf situations, the use of the turf will dictate the mowing height (Table 7-1). Sports turf managers generally mow fields at a lower cutting height to meet the demands of various games. Golf courses are classic examples of turf cut at extremely low heights. Homeowners and grounds managers can cut lawn and utility turfs at much higher mowing heights and still meet aesthetic purposes.

Turfgrass Species and Cultivar/Variety. Different species and, to a limited degree, some cultivars within a species have different mowing height tolerance ranges (Table 7-2). Creeping bentgrass (*Agrostis palustris*) and **hybrid** bermudagrass (*Cynodon dactylon*) can tolerate the low mowing height of a golf green; this same mowing height would destroy many other turfgrasses.

Many weeds such as annual bluegrass (*Poa annua*) and white clover (*Trifolium repens*) can tolerate lower mowing heights than can Kentucky bluegrass (*Poa pratensis*). To avoid a potential weed problem, the turf manager should match the mowing height to the grasses within a turf.

TABLE 7-1 Suggested Mowing Heights Based on Turf Use

Area	Cutting Height Range in. (cm)	Notes
Golf courses		
Greens	0.2–0.3 (0.5–0.8)	Some greens being cut below 0.2 in.
Tees	0.3–0.8 (0.8–2.0)	Influenced by species and play
Fairways	0.3–0.8 (0.8–2.0)	Influenced by play and species
Roughs	0.5–4.0 (1.3–10.0)	Influenced by play; shorter heights allow faster play
Athletic fields	0.5–2.5 (1.3–6.4)	Influenced by species, zone, and budget
Lawns		
Cool season	1.5–3.0 (3.8–7.6)	Higher heights encouraged for grass clipping recycling
Warm season	0.5–3.0 (1.3–7.6)	Influenced by species
Utility turfs	3.0–5.0 (7.6–12.7)	Many utility mowers limited in cutting height

Time of Season. Over the span of a growing season, changes in weather can result in unfavorable conditions for turfgrass growth. In many areas, poor weather can cause turfgrass plants to grow slowly or go into a dormancy. Many cool-season grasses go into a "summer dormancy" under the hot, dry weather conditions that are common in the summer.

When coming out of dormancy, a turfgrass plant depends on previously stored food (food reserves) to create new tissue. In particular, the plant produces new leaf tissue to begin the food production process again through **photosynthesis**. If these reserves are not adequate, the plant goes into stress, which results in reduced turf quality.

If the mowing height is raised before the onset of poor growing conditions, the plant has sufficient leaf tissue to develop food reserves from photosynthesis. For cool-season grasses, the turf manager should raise the cutting height at the onset of summer. For both cool-season and warm-season grasses, the turf manager should raise the height as colder weather approaches in the middle of autumn.

During periods of active growth and favorable weather, the turf manager can lower the mowing height to the lower end of the mowing height tolerance range. Lower mowing heights, however, will increase mowing frequency.

TABLE 7-2 Suggested Mowing Height Ranges of Turfgrass Species

Species	Height Range in. (cm)
Creeping Bentgrass	0.2–0.5 (0.5–1.3)
Kentucky Bluegrass	1.0–2.5 (2.5–6.4)
Perennial Ryegrass	1.0–2.5 (2.5–6.4)
Fine Fescue	1.0–2.5 (2.5–6.4)
Tall Fescue	1.5–3.0 (3.8–6.5)
Bermudagrass*	0.3–2.5 (0.8–6.4)
St. Augustine	1.5–3.0 (3.8–6.5)
Buffalograss	1.0–2.0 (2.5–5.1)
Centipedegrass	1.0–2.0 (2.5–5.1)
Zoysiagrass	0.5–1.0 (1.3–2.5)

* Varies depending on hybrid or cultivar.

Health of the Turf. When pests and traffic damage a turf, the turfgrass plant's only source of food for complete recovery is through photosynthesis. To speed recovery, the turf manager should raise the cutting height to increase total leaf surface area. This will enhance recovery through the increase in food from photosynthetic activity.

Mowing Frequency

The activity of mowing is an inherent stress on a turf. The distinguishing characteristic of a turf over other grasslands, however, is the regular and frequent mowing to which it is subject. A turf manager must cut a turf a regular intervals and with sufficient frequency to prevent the turf from appearing "unkempt." When determining the frequency of mowing, the turf manager must strike a balance between achieving a uniform appearance and placing excessive stress on the grass plants. It is interesting to note that in most situations, waiting too long between mowings can result in more damage to a turf than can mowing too frequently.

When developing a mowing program, the turf manager should be aware of the impact that mowing height has on the frequency of mowing. Waiting too long between mowings with short cutting heights will result in the removal of excessive amounts of leaf tissue and place the plant under increased stress. Routine mowing, such as the every-weekend mowing commonly practiced by homeowners, can result in potential pest problems and poor growth. To determine when to mow the turf, the turf manager or homeowner should follow the *one-third rule of mowing frequency*.

The one-third rule, in essence, states that to minimize stress on the turfgrass, mowing should not remove more than one-third of the leaf tissue at any one time. From a realistic standpoint, mowing should occur when the grass grows half-again (50 percent) higher than the desired cutting height (Figure 7-1).

If the grass becomes extremely tall because of delays in mowing, the turf manager should raise the cutting height to avoid violating the one-third rule. Then, over time, the turf manager can gradually lower the cutting height, always using care not to violate the one-third rule.

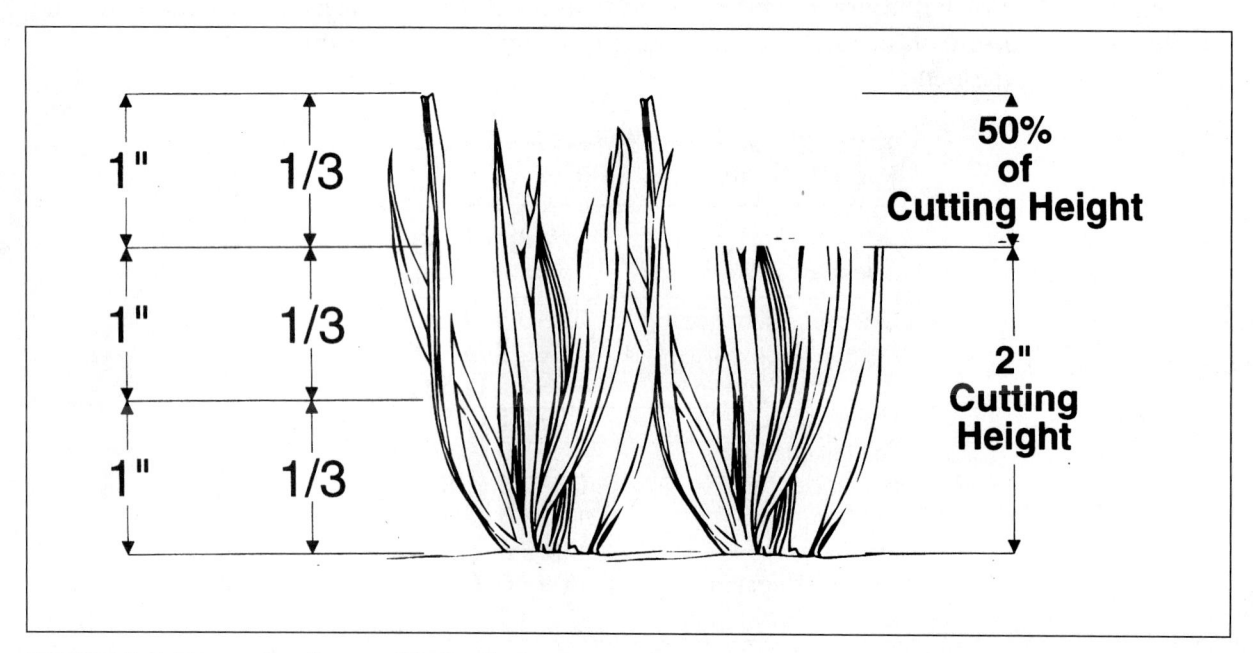

FIGURE 7-1 Example of one-third rule

MOWING EQUIPMENT

There are several different types of mowers available to the turf manager, including **reel mowers**, **rotary mowers**, flail mowers, and sickle-bar mowers.

Major Types of Mowers

Of the different types of mowers, the most prevalent are rotary and reel mowers. The other mowers are used primarily for utility turf mowing.

Reel Mowers. Turf managers use reel mowers on high-quality turfs that are maintained at low cutting heights. Golf courses, sports fields, and high-quality lawns consisting of bentgrass or hybrid bermudagrass look best when mowed with reel mowers.

Reel mowers cut with a scissors action, which leaves a clean edge on the leaf blade (Figure 7-2). This clean cutting action helps to maintain a quality surface feature which is important in high-quality turfs such as sports fields.

The actual mechanism of cutting consists of a series of blades forming a cylinder. Each blade is not parallel to the axis of the cylinder; each blade is at a spiral angle to axis.

As the cylinder or "reel" spins over a parallel-mounted **bedknife**, the blade and bedknife create a traveling cutting intersection similar in action to scissors. The position of the reel allows the reel blades to gather the grass leaf blades against the bedknife for cutting.

In relative comparison to other mowers, reel mowers require very little power to start the cutting action of the reels. By combining several reel units to form a *gang mower*, these large cutting swaths require very little additional power. The design of reel mowers also allows for very low cutting heights with little tearing of the grass. Golf course greensmowers can cut grass down to 1/8 inch (0.3 cm).

The reel mower is not without disadvantages. When mowing, reel mowers cannot easily cut tough seedhead stalks or tall grass above the axis of the reel. Also, reel mowers are precision cutting machines that require special maintenance to keep the cutting edges sharp. For large turf areas requiring a high quality of cut, such as golf courses and sod farms, however, the reel mower is the best choice.

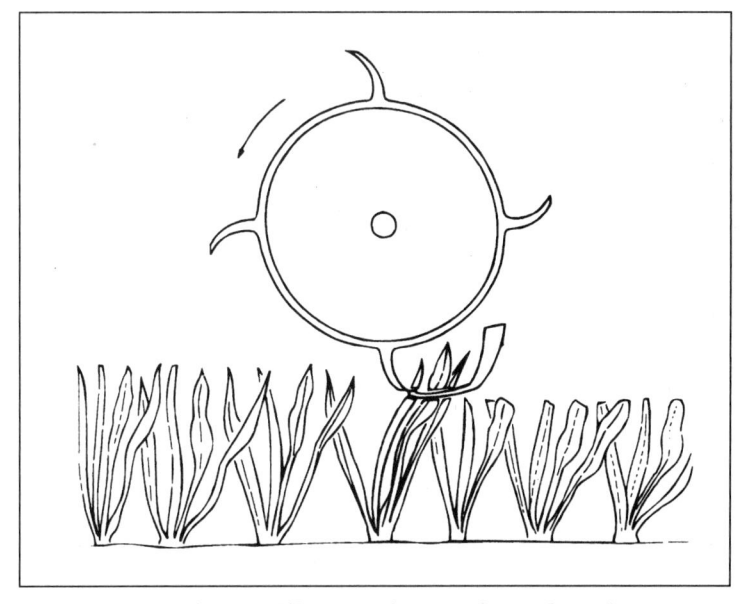

FIGURE 7-2 Scissors-like cutting action of reel mowers

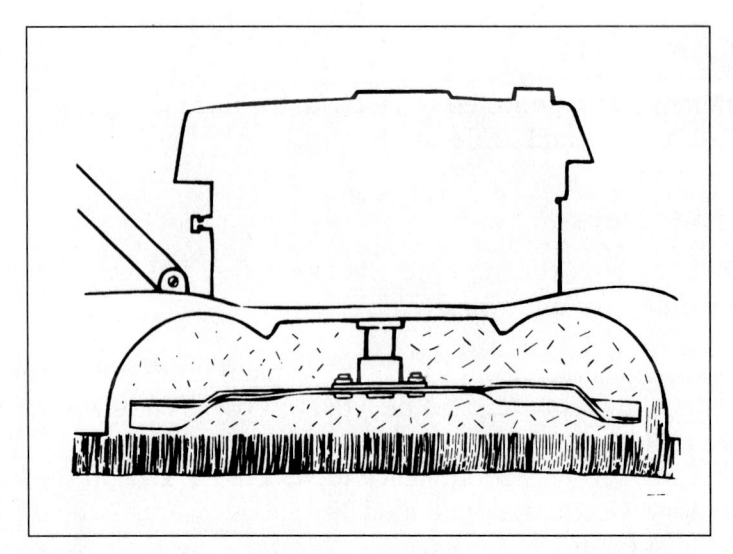

FIGURE 7-3 Chopping-like cutting action of rotary mowers

Rotary Mowers. Turf managers use rotary mowers for mowing lawns and many other types of turf where a moderate quality of cut at higher cutting heights is acceptable. These mowers cut grass by the impact of a horizontal blade rotating at a high speed (Figure 7-3). Unlike reel mowers, rotary mowers can cut tall grass and grass with tough seed stalks. Grounds managers often will use large rotaries to mow utility turfs because of the mower's ability to cut tall grass with seed stalks but still provide ease of maintenance (unlike flail mowers and sickle-bar mowers).

Maintenance of rotary mowers is relatively simple in comparison to that of reel mowers. Each blade has two knife-like edges that can easily be sharpened on a powered grinding wheel. The only additional requirement is to maintain a balanced blade during sharpening to avoid vibration damage to the equipment.

A major concern with rotary mowers is safety. The high speed blade can pick up rocks or other objects and eject them at high velocities. This can and many times will lead to injury of the mower operator or bystanders. Bystanders and operators must keep hands, feet, and clothing away from the operating mower deck to avoid potential loss of limbs.

Selecting a Mower

In addition to turf use and turfgrass species, the type of mower chosen for a turf will also depend on the following factors:

- the desired quality of cut
- mower operation and maintenance requirements
- labor availability and other management factors

Quality of Cut. A well-maintained reel mower will provide the highest quality of cut, particularly at very low mowing heights. For several grasses, such as creeping bentgrass, hybrid bermudagrass, and zoysiagrass, the only way to achieve a quality cut is with a reel mower. For most general purpose turf, a well-sharpened rotary mower will provide a very acceptable quality of cut.

Mower Operation and Maintenance Requirements. After selecting the type of cutting unit, the turf manager must look at the operation and maintenance requirements of the engine and the overall mower. Fuel and lubricant requirements, in particular, should be considered.

Mowers with diesel-fueled engines may be more efficient than mowers with gasoline-powered engines. The turf facility, however, may not be equipped to easily handle diesel-powered mowers, either from a fuel/fuel storage perspective or a service requirement perspective.

Several other maintenance factors such as drive belt servicing and blade sharpening will impact on mower selection. For a lawn maintenance company, drive belts that are hard to remove and replace could result in excessive downtime, inability to meet the customer's expectations, and loss of profits. For the same company, the inability to quickly change rotary blades in the field can result in similar consequences.

Operator safety and comfort are important factors related to safe mower operation. Uncomfortable mowers can lead to fatigue; this, in turn, can lead to fatigue related accidents and, potentially, to injury. When selecting a mower, the turf manager should consider the employee(s) who will be operating the mower.

Labor Availability and Other Management Practices. A major cost in any business operation is the employees. The industry trend towards larger, versatile mowers is to counteract the high cost of paying qualified operators. Superintendents have traditionally used walk-behind greensmowers to mow golf course greens. Several years ago, however, many superintendents switched to using larger triplex greens mowers. This reduced total mowing costs (labor, equipment, and overhead) by half even though the triplex greensmowers were three times the cost of equivalent walking greensmowers.

Aside from the costs of labor, some turf managers have problems finding skilled personnel to operate the mowing equipment. Many of today's commercial mowers are complex machines requiring a skilled operator.

MOWING COSTS

To develop a sound management plan and budget, the turf manager must determine reasonable estimates of the time required to mow any turf area. Many times, it is not feasible to run a series of tests to determine mowing times, particularly when the turf manager is shopping for a new mower during the midwinter trade shows. With certain data, however, mowing times can be estimated mathematically. The variables needed to estimate mowing times are:

- area of turf to be mowed
- effective cutting width (usually 85 percent of actual cutting width)
- speed of mower

The turf manager can then estimate mowing costs by using the following variables to extend out the calculations:

- labor costs per hour
- mowing frequency (times per week, month, or year)
- costs of the maintenance, gas, and oil
- any additional overhead expenses

The calculation for the time required to mow a given area is:

$$\frac{\text{area of turf}}{\text{effective cutting width}} = \textbf{distance traveled}$$

This calculation is similar to cutting the area into strips that are as wide as the effective cutting width of the mower, and attaching them end-to-end to form one long strip. The effective cutting width takes into account the small percentage of mower overlap.

After completing the calculation, one can use the result in a variation of the rate/time/distance formula:

$$\frac{\text{distance traveled}}{\text{speed of mower}} = \textbf{mowing time}$$

The turf manager can determine the cost of mowing by extending the mowing time calculation to include labor, fuel, and overhead costs.

There is a major flaw in the above calculation, however. The calculation does not take into account the time needed to turn mowers around when following a standard mowing pattern, that is the turn-around time. This is a variable factor ranging from 40 percent to 60 percent above the calculated time to mow a turf area. There are several factors that influence the turn-around times, including operator performance, the mower, the terrain, and mower maneuverability. Only from repeated field exercises can the turf manager determine a fairly accurate turn-around factor. For quick, fairly close estimates, 50 percent is a reasonable turn-around factor.

Example

A company buys a rotary mower for a five-acre corporate lawn area that will be mowed twice a week. The 72-inch mower has a mowing speed of 4 mph. How much should the grounds manager budget annually for mowing the lawn if labor costs average $12.00 per hour? (Assume an operating year of 40 weeks.) Note that the labor rate is the employer's total wage contribution; this includes expenses paid by the employer, such as taxes, social security, and insurance.

1. Convert the dimensions to consistent units such as feet.

   - area: 5 acres × 43,560 square feet per acre = 217,800 sq. ft.
   - width of 72-inch mower: $\frac{72 \text{ inches}}{12 \text{ inches per foot}} = 6 \text{ ft.}$
   - speed: 4 mph × 5,280 feet per mile = 21,120 fph (feet per hour)

2. Compute the distance traveled over the area.

$$\frac{\text{area}}{(\text{cutting width} \times .85)} = \textbf{distance traveled}$$

$$\frac{217,800}{(6 \times .85)} = \textbf{42,706 ft traveled}$$

3. Then determine the time required to mow an area once.

$$\frac{\text{distance traveled}}{\text{operating speed}} = \textbf{time required}$$

$$\frac{42{,}706}{21{,}120} = \textbf{2.0 hrs}$$

4. Account for time required to turn the mower around (turn-around factor of 50 percent).

$$\text{mowing time} \times (1 + \text{turning factor}) = \textbf{adjusted time}$$

$$2.0 \text{ hrs} \times (1 + 0.5) = \textbf{3.0 hrs}$$

5. Determine the total mowing time for 40 weeks per year.

$$\text{adjusted time of one mowing} \times \text{number mowings per week} \times \text{total weeks} = \textbf{total time}$$

$$3.0 \times 2 \times 40 = \textbf{240 hours of mowing}$$

6. Determine the total labor costs for the year.

$$\text{total time in hours} \times \text{labor rate} = \textbf{total labor cost}$$

$$240 \times 12 = \textbf{\$2{,}880.00 per year}$$

With a known fuel usage (MPG) and fuel cost, the annual fuel need can be determined from the distance traveled calculation:

$$\frac{42{,}706 \text{ ft in one mowing}}{5{,}280 \text{ feet per mile}} = \textbf{8.1 miles}$$

$$8.1 \text{ miles} \times 80 \text{ mowings} = \textbf{648 miles}$$

$$\frac{648 \text{ miles}}{4 \text{ MPG}} = \textbf{162 gallons}$$

$$162 \text{ gallons} \times \$1.10 \text{ per gallon} = \textbf{\$178.20 for fuel per year}$$

The calculation can be further extended using an overhead value to account for maintenance, depreciation, and other expenses.

ACTIVITIES

Operation of Different Mowers

With the guidance of the instructor or campus grounds manager, operate the different mowers used to mow the turf areas on campus. On the mowing operation and evaluation table (Figure 7-4) note various features such as the type of cutting unit, engine, and maneuverability.

	Mower 1	Mower 2	Mower 3	Mower 4	Mower 5
Type Cutting Unit (Rotary, Reel, Other)					
Engine Type					
Maneuverability					
Comments					

FIGURE 7-4 Mower operation and evaluation table

Trade Show or Distributor Field Trip

Attend a trade show or visit several mower distributors to see new mowers available on the market. If sales literature is available, collect one set of literature for each different type of mower (reel and rotary). Try to collect mower information different from that collected by your classmates.

Mowing Height Comparison

Mark out an evaluation area of turf on campus and mow sections of the turf at three heights. During the term or semester, evaluate the turf on a regular basis using standard quality evaluation ratings. On the mowing height evaluation table (Figure 7-5) note the incidence of pest problems, including a regular count of the number of weeds in each plot.

	2 Weeks		4 Weeks		6 Weeks		8 Weeks		10 Weeks	
	Quality Eval.	Weed Count	Quality Eval.	Weed Count	Quality Eval.	Weed Count	Quality Eval.	Weed Count	Quality Eval.	Weed Count
3/4"										
1.25"										
2.5"										

FIGURE 7-5 Mowing height evaluation table

Mowing Quality of the Two Major Types of Mowers

Equipment.
- reel mower
- rotary mower
- turf area

Procedure.
1. Mow half of a turf area with a reel mower and the other half of the same area with a rotary mower. Both mowers should be set at the same height.
2. Apply visual quality ratings to the mowed areas, comparing the effectiveness of the two mowers (Figure 7-6).

	Smoothness	Uniformity	Evenness of Cut
Reel			
Rotary			

FIGURE 7-6 Quality ratings (1 to 10) of rotary versus reel mower

Mower Speed Determination

The speed of any turf vehicle or machinery can be easily calculated without the use of a speedometer. In preparation for later activities, determine the speed of three mowers and note the results in the mower speed table (Figure 7-7).

Procedure.
Time the equipment under actual operating conditions.
1. Measure out 88 linear feet (1/60th of a mile).
2. Get the machinery up to operating speed. With a stopwatch, determine the time it takes for the equipment to travel the 88 ft distance.
3. Calculate speed in miles per hour (MPH) using the following formula:

$$\textbf{miles per hour} = \frac{60 \text{ seconds per minute}}{\text{seconds to travel 88 ft}}$$

	Mower 1	Mower 2	Mower 3
Time to Travel 88 ft (seconds)			
Speed (MPH)			

FIGURE 7-7 Mower speed table

Estimating Mowing Times

Using the table format in Figure 7-8 and the calculations presented in the discussion, calculate the amount of time required to mow an assigned turf area. If two or more mowers are available, calculate the estimated mowing times for all pieces of machinery.

Using Paper. Develop a paper copy of the tables to include in your record book along with the mower equipment records.

Using a Computer. Develop a spreadsheet of the above tables using a spreadsheet. You will need to add formulas to the appropriate cells.

On-Line Companion. U07A01.WKS—Mowing time estimates activity spreadsheet

	Mower 1	Mower 2	Mower 3
Area of Turf (sq ft)			
Mower Cutting Width (ft)			
Mower Overlap (%)			
Effective Cutting Width (ft)			
Mower Speed (MPH)			
Calculated Distance Traveled (ft)			
Calculated Time to Mow (hr)			
Turn-around Factor (%)			
Adjusted Time to Mow (hr)			

FIGURE 7-8 Estimated mowing time table

Field Comparison of Mowing Time Estimates

Equipment.
- measuring tape (100 ft)
- two stopwatches with start/stop capability
- calculator
- markers
- mowing machinery

After calculating all times in the previous activity, mow the assigned turf area to determine the accuracy of your calculations and note the results in the actual turf mowing times table (Figure 7-9).

	Mower 1	Mower 2	Mower 3
Estimated On-Turf Time to Mow Area			
Estimated Total Time to Mow Area			
Actual On-Turf Time to Mow Area			
Actual Total Time to Mow Area			
Percentage of Overall Time for Turns			

FIGURE 7-9 Actual turf mowing times table

Procedure.
1. Using a stopwatch with start/stop capability, record the time the mower is actually mowing on the turf area. The stopwatch should only be running when the mower is on the turf area. Stop the watch during turning-around of the mowers.
2. Using a second stopwatch, record the overall start to finish time required to mow the turf area. Do not include any stops for emptying grass clipping catchers.
3. Determine the percentage of overall time required to turn the mower around during cutting.

Estimating Mowing Cost for an Area of Turf

Using the information from the mower equipment file, the turf area file, the mowing times table, and additional information provided by your instructor, develop a table to determine the amount of time required to mow an assigned turf area once using the equipment as designated for each area. Account for turn-around time. Determine costs of mowing the campus one time. Consider costs to be $16.50 per mowing hour, which includes labor, gas, oil, and maintenance. After determining individual times, calculate an annual total cost.

Using Paper. Modify the paper copy of the tables to include in your three-ring binder along with the mower equipment records.

Using a Computer. Modify and enhance previously developed spreadsheets to perform the calculations. You will need to add formulas to the appropriate cells.

On-Line Companion. U07A02.WKS—Mowing cost activity spreadsheet

EXERCISES

Turf Mower Equipment File

Develop an equipment file of different mowers from several manufacturers. The record format should look similar to the sample mower equipment record (Figure 7-10).

As an additional exercise, add additional fields, such as original cost and purchase date, to your mower equipment file to enhance the information on the mowers.

Using Paper. Develop a paper form similar to the sample record (Figure 7-10). Make several copies of the form and develop a file of different mowers using literature obtained from trade shows and

82 Unit 7

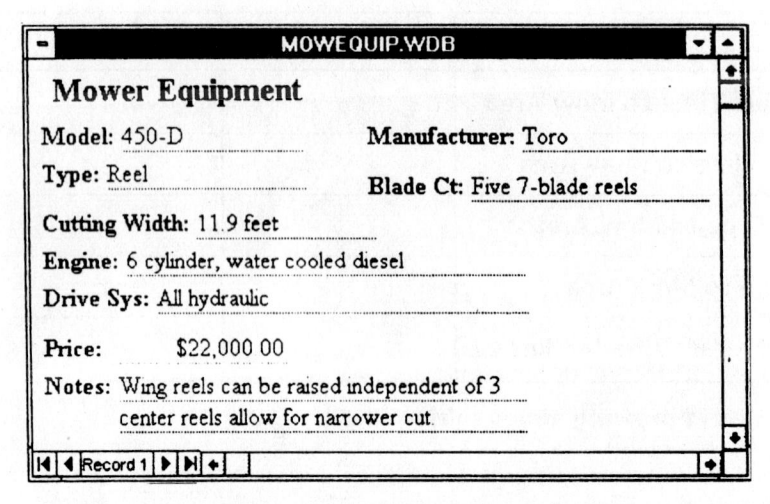

```
┌─────────────────────────────────────────────┐
│ −           MOWEQUIP.WDB              ▼ ▲   │
│                                          ↑  │
│  Mower Equipment                            │
│  Model: 450-D          Manufacturer: Toro   │
│  Type: Reel            Blade Ct: Five 7-blade reels │
│  Cutting Width: 11.9 feet                   │
│  Engine: 6 cylinder, water cooled diesel    │
│  Drive Sys: All hydraulic                   │
│  Price:    $22,000.00                       │
│  Notes: Wing reels can be raised independent of 3 │
│         center reels allow for narrower cut.│
│                                          ↓  │
│  |◄ ◄ Record 1 ► ►| ◄                       │
└─────────────────────────────────────────────┘
```

FIGURE 7-10 Sample mower equipment record

vendors as well as information found in trade journals. Include the file and the literature obtained from various sources in your three-ring binder.

Using a Computer. Develop a data file of different mowers using literature obtained from trade shows and vendors as well as information found in trade journals. Use the sample mower equipment record (Figure 7-10) as an example record layout. Include the literature obtained from various sources in your three-ring binder.

On-Line Companion. U07E01.WDB—Mower equipment database

Mowing Times and Cost Calculations and Records

Using Paper. Develop a table to estimate mowing times and costs for mowing the different areas of campus. Include in your three-ring binder along with your other records.

	B	C	D	E	F
2	Mower Width (in)	72			
3	Mower Speed (mph)	4.5			
4	% Overlap	15%			
5	% Turnaround	50%			
6	Labor per hour	$15.50			
7	Overhead/hour	$8.75			
8					
9	Campus Area	Lawns	Sports Field	Work Areas	Dorms
10	Area (sq. ft)	236,540	96,750	48,920	78,350
11	Mowings/Week	2	3	1	2
12	Weeks/Season	40	40	40	40
13	One time mowing (hr)	2.9	1.2	0.6	1.0
14	Time per season (hr)	234.2	143.7	24.2	77.6
15	Cost per season	$5,680.00	$3,485.00	$587.00	$1,882.00
16					
17				Total Time	480
18				Total Cost	$11,634

FIGURE 7-11 Sample mowing time/cost calculation form

Using a Computer. Develop a spreadsheet to estimate the mowing times and costs for mowing the different areas of campus (Figure 7-11). Use the information from previous exercises. As an alternative, if your spreadsheet program supports linking between different spreadsheets, create a new spreadsheet using the ability to link cells from other spreadsheets into your new spreadsheet.

On-Line Companion. U07E02.WKS—Mowing time and cost analysis spreadsheet

UNIT 8

Irrigation

Irrigation is the application of water to a turf. In many situations, there is not enough water supplied from natural precipitation to meet the current needs of the plant. Even during periods of normal precipitation, the survival of the turfgrass plant may depend on irrigation (such as during establishment).

Irrigation is also part of other cultural practices including the watering-in of chemicals such as fertilizers and **pre-emergent herbicides**. Under certain environmental conditions, the turf manager will run an irrigation system for a brief time (e.g., to cool the turf in hot summer or to remove frost on a cold morning).

DETERMINING THE NEED TO IRRIGATE

There are several methods that turf manager can use to determine the need to irrigate. Often two or more methods will be used to implement an appropriate irrigation practice for a turf area.

Visual Observation of Plant

As a plant goes into water deficit, close visual observation will reveal various changes to the plant. Caution must be practiced when using visual observation. Changes in the plant may be caused by things other than insufficient water in the soil. Changes noted through visual observation should, therefore, serve as indicators to use other methods of determining the need to water—before turning on the irrigation system.

Pre-wilting and Wilting Symptoms. As plants lose water, the turgor pressure of the leaves decreases. One indication of decreased turgor pressure and the potential for wilt is the appearance of footprints in the turf, which result from the inability of the leaves to spring back. Decreased turgor pressure is more noticeable in close-cut turf such as a golf green.

As a survival mechanism, turfgrass plants will start to roll, fold, curl, or twist leaves to reduce surface area exposure to drying conditions and, thus, retain water in the leaves. Bluegrass leaves will fold up along the midrib, while fine fescue leaves will roll up into thin tubes.

Under greater water deficits, the turfgrass plant will lose all turgor pressure and start to wilt; the leaves will go limp and start to shrink in size. If this wilting continues, leaf tissue will be lost and new leaves will have to grow from the crown.

Discoloration. Many turfgrasses will start to discolor as wilting starts to occur. The turf will usually take on a bluish-gray appearance. This color change is most noticeable with Kentucky bluegrass. As with other visual cues, however, discoloration can result from causes other than insufficient water, such as nutrient deficiency or chemical pollution.

Canopy Reflectance. An advanced tool for determining plant water need is measuring the infrared reflectance of turf. As plants go into stress such as water deficit, they absorb less and reflect more infrared light. Although infrared light is invisible to the human eye, there are several methods of detecting infrared reflectance. Methods of detection include infrared photography, canopy sensors, and computer imaging systems.

Soil Water Content

Because turfgrass plants obtain almost all their water from the soil through the roots, the water content of the soil is an important factor in determining the need to irrigate. When it comes to providing water to the turfgrass plant, the soil acts as a "storage tank." This tank constantly fills from natural precipitation and irrigation, while at the same time constantly empties through plant use, drainage, and evaporation. Even in winter, farmers look to winter snows and rains as a means of recharging the water table and improving soil moisture reserves for next season's crops.

The turf manager should know not only whether water is in the soil "tank"; but also the water holding capacity of the soil and how fast the soil will fill through rainfall or irrigation. Knowledge of these various factors enables the turf manager to develop an irrigation program that will result in little loss through runoff or drainage.

There are several methods the turf manager can use for determining soil water content. The turf manager may need to use two or more of the methods to obtain an accurate representation of soil water content.

Resistance Probes. Because water can act as a lubricant, as a soil dries, there is increasing resistance in the soil to penetration by a probe such as a screwdriver. With experience, water need can be determined to a moderate degree using this method. Although this method is not very accurate, it is a quick way to determine water need for localized watering purposes such as high areas of an undulating golf green.

Soil Color. A wet soil will generally have a darker color than a drier soil. By taking soil cores, the turf manager can quickly determine not only whether the soil contains water, but also the depth of wetness. A disadvantage of this method is the destruction of the turf in taking a sample.

Tensionmeters. As a soil dries, the soil will hold water under tension or negative metric potential because of the adhesive and cohesive nature of water molecules. A **tensionmeter**, also known as a potentionmeter, is a device for measuring soil moisture tension (Figure 8-1). It consists of a water-filled sealed tube with a porous ceramic cap. As tension increases in a soil due to decreasing water content, water in the tensionmeter will move through the ceramic cap until tension in the tube is equal to the soil moisture tension. If water is applied to the soil, thus reducing soil tension, water will move back into the tube through the cap until tension reaches equilibrium between the tube and the soil. A vacuum-type gauge indicates the tension in the tube, which directly relates to soil moisture tension.

Irrigation installers can connect tensionmeters to irrigation controllers, which switch irrigation on and off when conditions indicate a need for water. Tensionmeters have been used successfully in many turf situations, particularly high-quality athletic fields and golf greens. The use of tensionmeters is only practical, however, in fairly moist soils. Tensionmeters work best with specialty turfs, such as football fields and golf greens, where there is a need to maintain high soil moisture.

Electrical Conductivity. Although pure water does not conduct electricity, soil water, with its minute amounts of salt, will conduct an electrical current. As a soil becomes drier, there is a greater resistance to electrical conductivity. This resistance can be measured using various electrical instruments such as a volt-ohm meter. Because of the necessity for a minute amount of salt to conduct electricity, however, changes in the salt content resulting from fertilizers can cause an erroneous reading.

A common device for measuring electrical conductivity is the resistance block which consists of two wires imbedded in a porous material such as gypsum or a ceramic material. The blocks are permanently buried at various levels; the wires lead to a connector at the soil surface, and measurements are taken using a portable device. In order for this method to work effectively, the turf man-

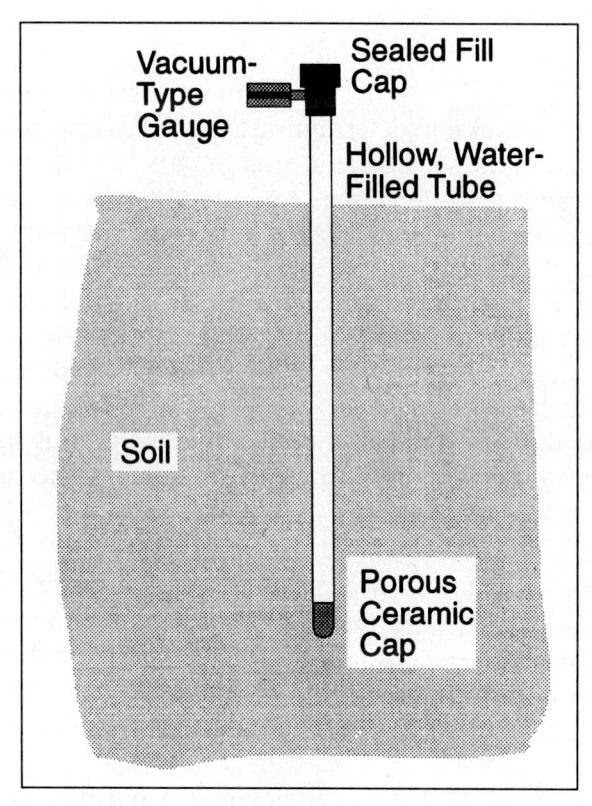

FIGURE 8-1 Tensionmeters are useful in measuring soil water level

ager must calibrate the readings to the actual soil moisture content for a given soil. This is done by using various advanced lab measurements to develop a calibration graph.

A resistance device commonly sold in garden centers consists of two metal probes emerging from a small, hand-held meter. Although not very accurate, these devices are useful in determining whether there is a need to irrigate a turf.

Advanced Lab Techniques. There are several advanced lab techniques, such as gravimetric sampling, tension tables, and pressure plates, used in determining soil moisture content. Most of the methods are destructive to a turf area and take a fair amount of time to obtain results. But they are useful and necessary in establishing graphs for use with the indirect soil moisture content measuring devices such as tensionmeters and resistance blocks.

Evapotranspiration Rates

When drainage water has left a soil, the two major causes of water loss are evaporation from the soil and **transpiration** from the turfgrass. These processes are related to each other in that they are both driven by the environmental conditions of wind, humidity, sunlight, and temperature. Soil scientists combine evaporation and transpiration into a process known as **evapotranspiration** or **ET**. Knowledge of the ET rate as an indicator of water loss is important in irrigation scheduling. The ET rate helps the turf manager in determining how much water to apply to a dry turf area.

A common method for determining ET rates is to fill a pan with water and measure the amount of water lose through evaporation. The amount of water lost correlates to the ET rate of a particular crop. Crop scientists define the relationship between pan evaporation and ET rates of each crop as the *Crop Coefficient* (Kp). For turf purposes, turfgrasses are considered various kinds of "crops," each with a different crop coefficient. Several other devices are also available for measuring ET rates.

Evapotranspiration rates are available from several sources including university Extension services, state weather networks, and municipal water departments. For large turf areas such as golf courses, a superintendent can measure ET rates using instruments available with on-site weather stations.

Weather Stations, Records, and Forecasts

Use of various weather information sources can go a long way in developing an efficient irrigation program. Even learning to effectively read the barometer of an on-site weather station can help the turf manager in determining whether to operate the irrigation system.

Weather records are useful for determining when various weather events are likely to occur during the growing season. A good example is the regular droughts that occur in many areas each summer. By anticipating these droughts, the turf manager can perform the necessary service on an irrigation system, thus avoiding costly breakdown of equipment and the resulting loss of turf during the drought.

Even though weather forecasts are not 100 percent accurate, in most instances they are accurate enough to be used in anticipating the need to irrigate. Weather forecasts are now available through a variety of means, including radio, television, and the Internet.

For many high-maintenance turf sites, particularly golf courses, the turf manager takes on the role of meteorologist. The turf managers, with the aid of a computer and modem, can now access the same weather satellite and radar images as do professional meteorologists at the airport or nearby television studio. Many of the images are available on a near real-time basis so that the turf manager can access the images within minutes after weather satellite transmission.

IRRIGATION SCHEDULING

In many areas, water is a limited and precious resource that the turf manager cannot afford to waste. In many localities, irrigating a turf area in a manner resulting in water runoff can lead to several penalties.

Determining the amount of water to apply has traditionally involved a fair amount of guesswork; the turf manager or homeowner would set the irrigation timers hoping to apply a sufficient amount of water to meet the needs of the plant. Often the only determination of water need was through visual observations. From cultural, environmental, and financial standpoints, water is a valuable asset that the turf manager must carefully handle like any other asset.

Water Budgeting

The *water balance*, or *water budget*, method is a simple way to schedule irrigation. This method takes into account several known and predictable factors to produce an irrigation schedule that will meet the needs of the turfgrass while reducing waste.

The *checkbook method* of irrigation scheduling is another way to budget water. Irrigation and rainfall are considered checkbook deposits, while evapotranspiration and deep percolation are considered checkbook withdrawals. By monitoring deposits (of irrigation and rainfall) and withdrawals (from evapotranspiration), the turf manager always knows the checkbook balance (of water). By maintaining all factors based on water depth, it is simple to keep a balanced water checkbook.

A water-budget worksheet (Figure 8-2) can be set up for use with the checkbook method. (Either columnar paper and a calculator or a computer spreadsheet (Figure 8-3) can serve as a water-budget worksheet.)

Date	Initial Water Content	Plus Effective Rainfall	Plus Irrigation Depth	Minus Evapo-trans-piration	Minus Deep Perco-lation	Final Water Content
5/30	3.00	0.00	0.00	0.29	0.00	2.71
6/5	2.71	0.00	0.00	0.28	0.00	2.43
6/11	2.43	0.10	0.00	0.35	0.00	2.18
6/16	2.18	0.00	0.75	0.32	0.00	2.61
6/21	2.61	3.55	0.00	0.08	2.50	3.58

FIGURE 8-2 Sample water budget worksheet

Setting up a water budget requires knowledge of the following factors (in inches or centimeters):

- water content at beginning and end of the time period
- effective rainfall
- irrigation depth
- evapotranspiration rate
- deep percolation

NUMBER	DATE	CHECKS ISSUED TO OR DESCRIPTION OF DEPOSIT	(−) AMOUNT OF CHECK	(+) AMOUNT OF DEPOSIT	BALANCE
	5/30	Initial Water Content		3.00	3.00
					3.00
	5/30	Evapotranspiration	0.29		0.29
					2.71
	6/5	Evapotranspiration	0.28		0.28
					2.43
	6/11	Effective Rainfall		0.10	0.10
		Actual: .34"			2.53
	6/11	Evapotranspiration	0.35		0.35
					2.18
	6/16	Irrigation		0.75	0.75
					2.93
	6/16	Evapotranspiration	0.32		0.32
					2.61
	6/21	Rainfall		3.55	3.55
		Very Heavy Rainfall on 6/18-20			6.16
	6/21	Evapotranspiration	0.08		0.08
		Very Cool and Cloudy			6.08
	6/21	Deep Percolation	2.50		2.50
					3.58

FIGURE 8-3 Water budget in checkbook ledger format

In addition to these factors, one must determine the available water content in the soil. Plant-available water content is the water between field capacity (no more drainage water) and the wilting point (when the plant can no longer extract water). From the plant-available water factor, the turf manager assigns an allowable depletion factor. This factor is the amount of water depletion that results in an acceptable stress on the plant.

Continuing the checkbook analogy, the allowable depletion factor is the minimum balance that must be maintained to avoid a penalty. As with some bank checking accounts, when the balance goes below a minimum level, the bank applies service charges to the account. The stress penalties on a plant, such as when the water content goes below one inch, are the "service charges" deducted from the account.

Water Content. Water content is measured in inches and can be determined using a knowledge of soil texture, soil structure, and rooting depth. Many soil labs can test the soil for water content parameters. Using a calibrated soil-moisture meter, the turf manager can easily measure the soil water content at any time.

Effective Rainfall. Effective rainfall is rainfall that enters the soil profile minus any rainfall lost due to runoff. Some agricultural weather reporting stations give effective rainfall values. By observing rainfall intensity and runoff, the turf manager can approximate estimates of effective rainfall. With a knowledge of the soil infiltration rate and percentage slope of the turf, the turf manager can determine a more accurate effective rainfall value.

Irrigation Depth. The amount of water applied with an irrigation system is measured as a depth over an area. One can calculate the amount of water applied from the precipitation rate of a sprinkler system and the sprinkler run time. Precipitation rate is measured in inches per hour and is calculated using the following formula:

$$\frac{\text{Total gallons per minute of area} \times 96.3}{\text{area irrigated in square feet}} = \textbf{precipitation rate}$$

When the precipitation rate is known, the irrigation run time required to deliver a given amount of water can be calculated:

$$\frac{\text{water application in inches}}{\text{precipitation rate in inches per hour}} = \textbf{irrigation run time}$$

Conversely, one can determine a total amount of applied water from the irrigation run time along with the precipitation rate.

$$\text{irrigation run time in hours} \times \text{precipitation rate} = \textbf{total water applied}$$

Evapotranspiration. As discussed earlier in the unit, ET is the rate of water loss from both soil evaporation and plant transpiration. This rate is available from several sources. Many resources also provide potential ET rates based on past records. The turf manager can use these potential rates in the checkbook for advanced irrigation planning. With computer spreadsheets, several "what-if" scenarios can be studied to plan irrigation programs in advance.

Deep Percolation. Water that moves beyond the root zone during downward percolation is water lost to the plant. The turf manager can estimate deep percolation using various means including soil texture, infiltration rates, percolation rates, and soil moisture measuring devices. To minimize deep percolation, the turf manager should encourage deep rooting of the turfgrass (Figure 8-4).

FIGURE 8-4 Deep rooting minimizes loss by deep percolation

UNIFORM WATER APPLICATION

The primary goal in the designing, using, and installing a turf irrigation system is the uniform application of water to the turf at the proper rate and timing. In practice, sprinkler distribution patterns are not perfectly uniform. Factors influencing the distribution pattern of sprinkler systems include wind, head spacing, nozzle pressure, rotation, nozzles, risers, age of system, and water condition. Because of these factors, the turf manager should occasionally check the sprinkler distribution patterns over important areas of irrigated turf to ensure uniform applications of water.

The uniformity of a given irrigation distribution pattern can be estimated by testing for a uniformity percentage (u%). The uniformity percentage is determined by entering the test results into the calculation:

$$U\% = 100 \times 100 (1.0 - [\text{sum of differences} / \text{total volume}])$$

where:

sum of differences = the sum of the absolute values of the individual amounts of water collected in each container subtracted from the average amount of water collected in each container, and

total volume = Total volume of all individual containers

Determination of uniformity involves placing collection containers of a uniform size and shape, such as coffee cans, on an irrigated area in a grid or in a random fashion. The position of each collection container should be recorded to identify malfunctioning sprinklers after the test. The area is irrigated until the containers have a measurable quantity of water. Following this, the cans are collected, the water volume of each container is measured, and the data is recorded. Processing the test results in the given calculation provides the uniformity percentage (Figure 8-5). The uniformity percentage should be greater than 85 percent for a well-designed, functioning sprinkler system.

Can Number	Volume (ml)	Average Volume (ml)	Average Values (ml)
A1	170	167.2	2.8
A2	150	167.2	17.2
A3	195	167.2	27.8
B1	191	167.2	23.8
B2	141	167.2	26.2
B3	155	167.2	12.2
C1	190	167.2	22.8
C2	163	167.2	4.2
C3	158	167.2	9.2
D1	191	167.2	23.8
D2	154	167.2	13.2
D3	149	167.2	18.2
Total Volume:	2007 ml	Sum of Differences:	201.4 ml

U% = 100% * (1.0 − (Sum of Differences / Total Volume))
= 100% * (1.0 − (201.4 ml / 2007 ml))
= 90.0%

FIGURE 8-5 Sample irrigation uniformity percentage calculation

ACTIVITIES

Soil Water Content

Determine the soil water content of various turf soils using the gravimetric determination method.

Equipment.

- greenscup cutter, soil core extractor, or bulb planter with uniform diameter
- balance
- knife
- ruler
- drying oven

Procedure.

1. Using the cup cutter, soil core extractor or bulb planter, remove a soil core from a turf area. Do not allow the soil core to break apart. Attempts should be made to keep the core as undisturbed as possible.
2. With the knife, cut off the vegetation and thatch layer to the soil surface.
3. Using the ruler as a measuring guide, cut a uniform-sized sample from the soil core. All samples should be the same size and from the same location of the core.
4. Immediately weigh the soil sample. Use metric values. Record the wet-soil-sample weight. If there is an extended time between sampling and weighing, store the sample in a container to prevent drying.
5. Dry the sample in the oven for twenty-four hours at 105°C (221°F).

6. Weigh the dry sample. The difference between the wet-soil weight and the dry-soil weight is the weight of water. Record the dry-soil weight and water weight. Subtract out the weight of any retainer ring used to hold the sample together.
7. Using the ruler, measure the soil sample and calculate the volume of the sample. Use metric measurements. Record this value and calculate the bulk density (grams/cm^3) using the fomula:

bulk density = weight of soil/volume of soil

8. Calculate the water content by volume using the formula:

water content by volume as % = (weight of water/weight of soil) × bulk density

Soil Moisture Measurements

Work with various soil moisture measuring methods, noting advantages and disadvantages of each.

Equipment.
- tensionmeter
- resistance blocks

Insert these devices into several different soils and take measurements over time. Note changes in soil moisture content.

Soil Moisture Calibration Curve

Calibrate the sensors to the soil by determining soil moisture contents using gravimetric methods. Develop a calibration graph to convert sensor readings to soil moisture content.

Procedure.
1. Insert one or more tensionmeters and resistance blocks into a turf soil tested in the Soil Water Content activity.
2. Irrigate the soil until saturated.
3. Immediately take readings from the sensors and determine water content using gravimetric methods. Record all data. *Note:* Being that bulk density has already been determined (in the Soil Water Content activity), the soil samples can be distributed during collection.
4. At regular intervals as designated by your instructor, take readings from the sensors and determine water content using gravimetric methods. Record all data.
5. Using the data, plot a curve with water content by volume on the *x* axis and sensor readings on the *y* axis. A computer graphing program may be used in plotting the curve.

Save this soil moisture calibration curve for use in the Water Budget exercise.

Evapotranspiration Rates

Observe different methods for determining ET rates. Use various resources for obtaining ET rates, including the state Extension Service and agriculture information networks.

Accessing Weather Records and Forecasts

Compile a list of the different means of accessing weather information and forecasts, including computer networks and bulletin boards. Keep a log of the following weather information:
- high and low temperature
- precipitation

- average humidity
- average wind speed
- evapotranspiration rate
- percent cloud cover

Sprinkler Application Uniformity

Determine the uniformity percentage for one or more areas of irrigated turf as designated by the instructor. If the uniformity percentage is lower than 85 percent, note why it is lower than expected.

Procedure.
1. Place twelve cans 5 to 10 feet apart in a uniform or random pattern in the test area of turf. Stake the cans if necessary to prevent tip-over from wind.
2. Irrigate until the cans contain a measurable amount of water.
3. Measure and record the amount of water collected from each container in the irrigation uniformity percentage calculation table (Figure 8-5).
4. Determine the uniformity percentage using the calculation method described earlier in the unit. Write the result in the space provided in the uniformity data tables (Figure 8-6).

Area:			
Can Number	Volume	Average Volume	Absolute Values
Total Volume:		Sum of Differences:	
Uniformity Percentage:			

Area:			
Can Number	Volume	Average Volume	Absolute Values
Total Volume:		Sum of Differences:	
Uniformity Percentage:			

Area:			
Can Number	Volume	Average Volume	Absolute Values
Total Volume:		Sum of Differences:	
Uniformity Percentage:			

Area:			
Can Number	Volume	Average Volume	Absolute Values
Total Volume:		Sum of Differences:	
Uniformity Percentage:			

FIGURE 8-6 Irrigation uniformity data tables

Using a Computer. Develop a spreadsheet of Figure 8-6.

On-Line Companion. U08A01.WKS—Irrigation uniformity activity spreadsheet

Irrigation Scheduling

Using the information from previous activities (including those in activities in previous units), schedule the campus irrigation system to maintain a water balance using the checkbook method. Use a copy of the worksheet shown in Figure 8-3.

EXERCISES

Soil Moisture Log

With tensionmeters or resistance blocks, keep a log of soil moisture content at all three different zone depths (3 inches, 6 inches, and 12 inches) over time. Plot a chart of the soil moisture content over time.

Using Paper. Develop a log-sheet form and maintain the logs using the form. Use graph paper to plot the soil moisture content over time. Keep all forms and charts in your three-ring binder.

Using a Computer. Develop the logs using either a spreadsheet or database. Use the charting feature built into the program or an external charting program to create the soil moisture content calibration chart developed in the Soil Moisture Calculation Curve activity for use with the sensors. In addition, create a chart of soil moisture content for various root depths over time.

On-Line Companion. U08E01A.WKS—Soil moisture log spreadsheet
U08E01B.WDB—Soil moisture log database

Weather and Evaporation Rate Logs

Maintain a log of weather conditions and ET rates during the entire period of your turfgrass management course. Keep track of the conditions listed in the previous activity.

Using Paper. Develop a log sheet form and maintain the logs using the form. Use graph paper to plot the weather conditions over time. Keep all forms and charts in your three-ring binder.

Using a Computer. Develop the logs using either a spreadsheet or database. Use the charting feature built into the program or an external charting program to plot the various conditions over time.

On-Line Companion. U08E02A.WKS—Weather record spreadsheet
U08E02B.WDB—Weather record database

Water Budget Checkbook

Using Paper. Develop a water-budget checkbook ledger using either columnar paper or an unused bank-check register. The layout should make it easy to calculate and maintain a checkbook balance. Include the ability to calculate irrigation run time based on sprinkler system precipitation rates and the amount of irrigation depth needed to maintain the balance.

Using a Computer. Develop a water-budget checkbook using either a spreadsheet or database. It should include the appropriate formulas to maintain the necessary checkbook balance. Include the ability to calculate irrigation run time based on sprinkler system precipitation rates and the amount of irrigation depth needed to maintain the balance.

On-Line Companion. U08E03A.WKS—Water-budget checkbook spreadsheet
U08E03B.WDB—Water-budget checkbook database

UNIT 9

Turf Pest Problems

Pests are organisms, both plants and animals, that reduce turf quality by feeding on the turfgrass plants or competing for the same growth factors as those needed by turfgrass plants. Pest problems usually come in one of three major varieties: *weeds, insects,* or *disease.*

Which pest problem is of primary concern varies depending on the use of the turf, the quality level of the turf, and turfgrass management practices. For homeowners, the biggest concern is often weeds, followed, in turn, by insect problems (usually grubs) and disease. Due to the stress placed on golf turf and the importance of a quality putting surface, disease is often of prime concern to the golf course superintendent.

Identification and understanding of turf pest problems allow the turf manager to establish appropriate controls. Given the continuing concern regarding the effects of pesticides on the environment, the turf manager needs to work toward not only reducing the pest problem, but also correcting the underlying cause of the problem.

TURF WEEDS

Any plant growing where one does not want it is a weed. A plant becomes a weed in the turf when it disrupts the uniformity of the turf and competes with desirable turfgrass species for growth factors including light, water, and nutrients. A plant may, however, be a weed in one situation and a desirable plant in another situation. For example, creeping bentgrass, the most widely used turfgrass species on golf greens in the northern United States, is often a weed in well-maintained Kentucky-bluegrass lawns.

Weeds are classified as *broadleaf, grassy,* or *sedges* (Figure 9-1). Most weeds are broadleaf or grassy; a few are sedges. Broadleaf weeds are also called **dicot** weeds—short for **dicotyledon**, or two-seed leaves. Grassy weeds are also called **monocots** (short for **monocotyledon**, or one-seed leaf), which are in the same class as turfgrass plants. Weeds can also be classified by growth cycle as **annuals**, **biennial**, or **perennial**.

An annual completes a full life cycle, from seed to seed, in one growing season. During the nongrowing season, the plant dies, leaving only seeds for the next growing period.

A biennial completes a full life cycle in two growing seasons. After a season of vegetative growth, the plant goes into dormancy during the first nongrowing period. In the second growing season, it goes into a reproductive phase and then dies, leaving seed for the next cycle.

A perennial is capable of surviving three or more seasons. New growth occurs each season from vegetative tissue that survives during the nongrowing season. Because weeds are often herbaceous (meaning the top-growth dies back during the nongrowing season), new growth occurs each season from underground structures such as rhizomes and bulbs.

Perennial weeds also go through a reproductive phase, wherein they produce new seed. Unlike annuals and biennials, these plants do not completely die at the end of a growing season. Rather, the plants go back into a period of vegetative growth.

Identification of common turfgrass weeds (Table 9-1) and their growth cycles allows for the implementation of appropriate control measures. For example, the turf manager can control crab-

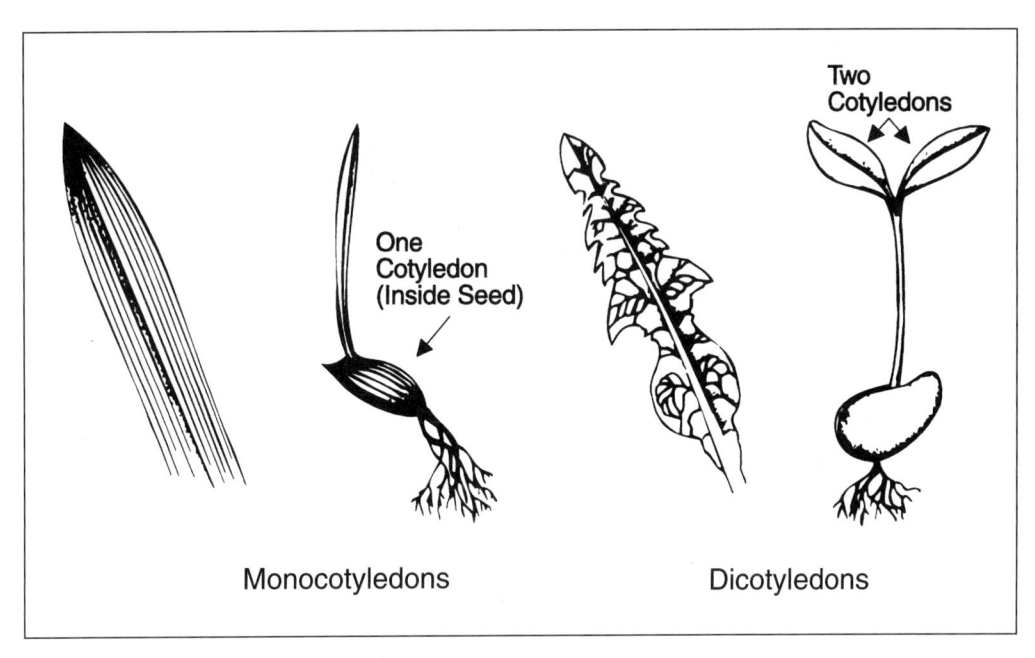

FIGURE 9-1 Comparison of grassy weeds and broadleaf weeds

TABLE 9-1 Common Turfgrass Weeds

Annual Grasses	Perennial Grasses	Broadleaf Weeds		
Bluegrass, annual	Bentgrass	Binweed, field	Garlic, wild	Pigweed
Barnyardgrass	Bermudagrass	Burdock	Henbit	Plantain, broadleaf
Crabgrass, hairy	Tall Fescue	Carpetweed	Ivy, ground	Plantain, buckhorn
Crabgrass, smooth	Johnsongrass	Carrot, wild	Knotweed	Puslane
Foxtail, green	Kikuyugrass	Chickweed, common	Lambsquarter	Shepherdspurse
Foxtail, yellow	Knotgrass	Chickweed, mouse-ear	Mallow	Speedwells
Goosegrass	Nimblewill	Cinquefoil	Medic, black	Spurges
Panicum, fall	Nutsedge, yellow	Clover, white	Mustard, wild	Thistles
Rescuegrass	Quackgrass	Dandelion	Onion, wild	Violets, wild
Sandbur	Velvetgrass	Dock, curly	Pennywort	Yarrow, common

grass, a common annual grassy weed, with a pre-emergent herbicide in the spring as the seeds germinate. Using herbicides on crabgrass in late summer, however, is a waste of material because the crabgrass plant will die with the onset of cold weather.

TURF INSECTS AND RELATED PESTS

Insects are small animals ranging from nearly microscopic to about 6 inches (15.24 cm) in size, with six legs attached to a three-segmented exoskeleton encased body. Other characteristics include antennae and compound eyes (Figure 9-2). Insects feed on the grass plants, resulting in severe damage and possible loss of the turfgrass plant. Out of the over one million species of insects, only a relatively small number cause damage to turf (Table 9-2). Most insects do not harm a turf, and numerous species are beneficial to turf ecology.

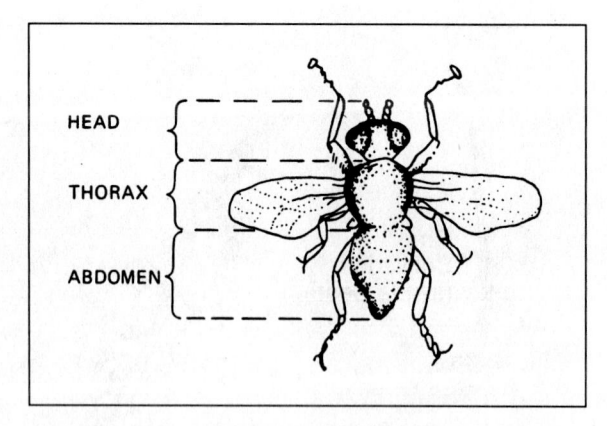

FIGURE 9-2 Major structures of an insect

TABLE 9-2 Common Turf Insect Problems

Common Name	Technical Name
Soil Inhabitants	
Black turfgrass atenius	*Ataenius spretulus*
Bluegrass billbugs	*Sphenophorus* spp.
Ground pearls	*Margarodes meridionalis*
Mole crickets	*Scapteriscus acletus*
Grubs	
Japanese beetles	*Popilla japonica*
June beetles	*Phyllophaga* spp.
Northern masked chafers	*Cyclocephala borealis*
Southern masked chafers	*Cyclocephala immaculata*
Thatch Inhabitants	
Black cutworms	*Agrotis ipsilon*
Bronzed cutworms	*Nephelodes minians*
Fall armyworms	*Spodoptera frugiperda*
Hairy cinchbugs	*Blissus leucopterus*
Hyperodes weevils	*Hyperodes* spp.
Sod webworms	*Crambus* spp.
Southern cinchbugs	*Blissus insularis*
Stem and Leaf Inhabitants	
Bermudagrass mites	*Eriphyes cynodoniensis*
Bermudagrass scales	*Odonaspis ruthae*
Clover mites	*Bryobia praetiosa*
Rhodesgrass scales	*Antonina graminis*
Winter grain mites	*Penthaleus major*

Mites

Mites are eight-legged, two-segmented animals that are similar to insects but belong to the arachnid, or spider family. Although they are not insects, mites can cause similar damage to a turf. In comparison to many turf insect pests, mites are a minor problem. Due to differences in physiology, however, insecticides may not control mites. Chemical control may require the use of miticide to specifically control a mite pest problem.

The Life Cycle of an Insect

As an insect hatches from an egg and grows, it goes through several stages in its life cycle. As an insect progresses from one stage to another, it often changes in shape or form. This is called **metamorphosis**.

Insects that change drastically in appearance from egg to adult go through a *complete metamorphosis* (Figure 9-3). The four major stages of complete metamorphosis are egg, larva, pupa, and adult. At the larval stage, an insect is often quite different than at the adult stage. The change from larva to adult occurs during the pupal or "resting" stage.

When an immature insect (**nymph**) looks similar to the adult in all respects except size, the insect has gone through a simple metamorphosis (Figure 9-4). As the immature insect sheds the old exoskeleton through molting to a new instar, it gradually changes in size and characateristics into an adult.

A knowledge of the life cycle of an insect is important in anticipating potential damage and in controlling insect pest problems.

Some insects that undergo complete metamorphoses are most damaging to a turf during the larval stage and cause little or no damage during the adult stage. For example, the sod webworm is the damaging larval stage of the innocuous lawn moth. With other insect species, such as the Japanese beetle, the insect causes damage during both the larval and the adult stages.

In many situations, it is much easier to control the pest in one stage as opposed to another. For example, for insects that undergo a simple metamorphosis, such as the chinch bug, the earlier immature instar is much easier to control because it is more susceptible to lower toxin levels than are later instars or adults.

Feeding Mechanisms and Feeding Locations of Insects

In addition to having knowledge of the life cycle, the turf manager must be familiar with the feeding mechanisms of the insect pests, as well as the feeding locations of the pests. Some insects chew

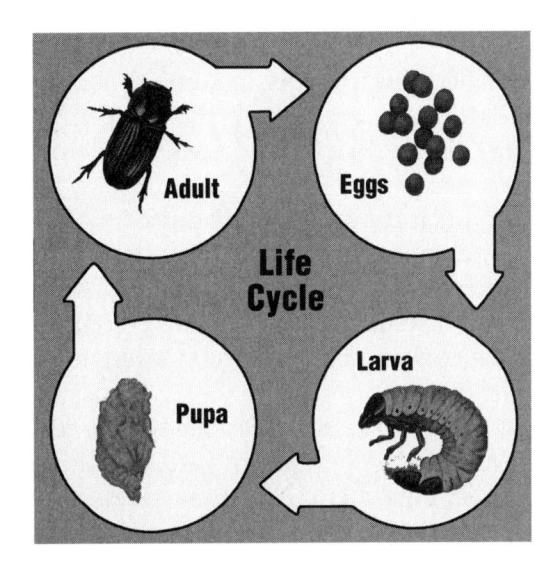

FIGURE 9-3 Example of complete metamorphosis

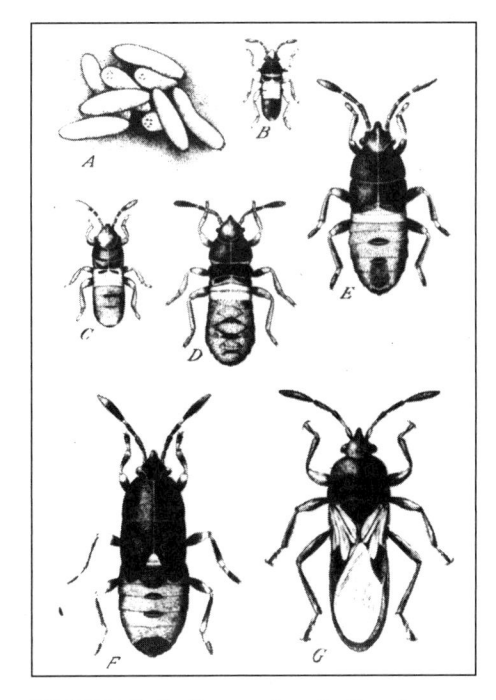

FIGURE 9-4 Example of simple metamorphosis

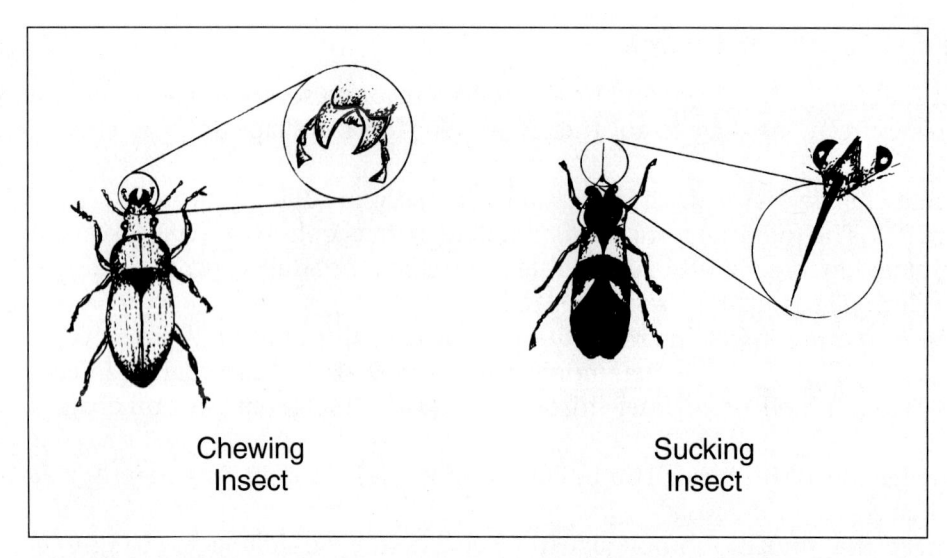

FIGURE 9-5 Typical chewing and piercing-sucking mouthparts

plant tissue using jaw-like mouthparts. Other insects have piercing-sucking mouthparts for drawing plant liquids from leaf, stem, or root tissue (Figure 9-5). Sucking insects are often harder to control because their thin, needle-like mouthpart can potentially bypass any insecticide sitting on the surface of a leaf.

Different insect pests feed on different parts of a turfgrass plant. Grubs are root-feeding pests and will not likely be observed on the grass leaves. Knowledge of where the pest feeds helps in finding the pest and, thus, determining the existence of an insect pest problem.

Feeding location is important in chemical control of insect pests because the goal is to place the chemical where the pest feeds on the plant. For leaf-feeding insects, for example, the chemical must be placed on the leaves. Rainfall after application of a chemical for controlling leaf feeders can reduce the effectiveness of the chemical.

TURF DISEASES

A disease is a plant disorder caused by microorganisms called **pathogens**. In some situations, disease is due to nonpathogenic agents such as pollution. Pathogens can be fungi, bacteria, viruses, or **nematodes**. Various species of fungi cause the major turf diseases; bacteria, viruses, and nematodes cause a few minor diseases (Table 9-3).

In order for a disease to develop in a plant, several factors must occur at the same time. A plant must be susceptible to the pathogen and be in the proper physiological (i.e., "host") state. The disease organism must be present in proximity to the plant. In addition, environmental conditions must be favorable for pathogen growth and infection. The factors of host plant, pathogen, and environment create the disease triangle (Figure 9-6). If one of the three conditions is missing, a triangle cannot form, and a disease will not occur in the plant.

Several factors cause a plant to be in a host condition for a pathogen. Some grass species and varieties are just more susceptible to certain diseases than are others. Use of resistant varieties and species in a turf stand can help considerably in controlling disease. Improper cultural practices, including under- and over-fertilization, can create a physiological condition that makes the plant more susceptible to **infection** by a pathogen. Other stresses on the turf, such as improper mowing, also can put the plant in a favorable condition for attack from the disease pathogen.

TABLE 9-3 Common Turf Diseases

Disease	Pathogen
Anthracnose	*Colletotrichum graminicola*
Brown patch	*Rhizoctonia solani, R. zeae*
Copper spot	*Gloeocercospora sorghi*
Dollar spot	*Scleortinia homeocarpa*
Fairy ring	*Marasmius oreades* and others
Gray leaf spot	*Pyricularia grisea*
Gray snow mold	*Typhula* spp
Leaf spots, blights	*Drechslera* spp and others
Necrotic ring spot	*Leptosphaeria korrae*
Pink path	*Limonomyces roseipellis*
Pink snow mold	*Microdochium nivale*
Powdery mildew	*Erysiphe graminis*
Red thread	*Laetisaria fuciformis*
Rusts	*Puccinia* spp
Slime mold	*Plasmodium*
Southern blight	*Sclerotium rolfsii*
Spring dead spot	*Leptosphaeria korrae*
Summer patch	*Magnaporthe poae*
Smut, stripe	*Ustilago striiformis*
Take-all patch	*Gaeumannomyces graminis*
Yellow tuft	*Scherophthora macrospora*

With most turfgrass diseases, the pathogen is already present in the turf. The pathogen either is dormant or lives on dead organic matter until conditions are favorable for it to infect a living grass plant. Under active conditions, some pathogens spread into noninfected turf via equipment or water movement. Sanitation can therefore help to contain some disease problems.

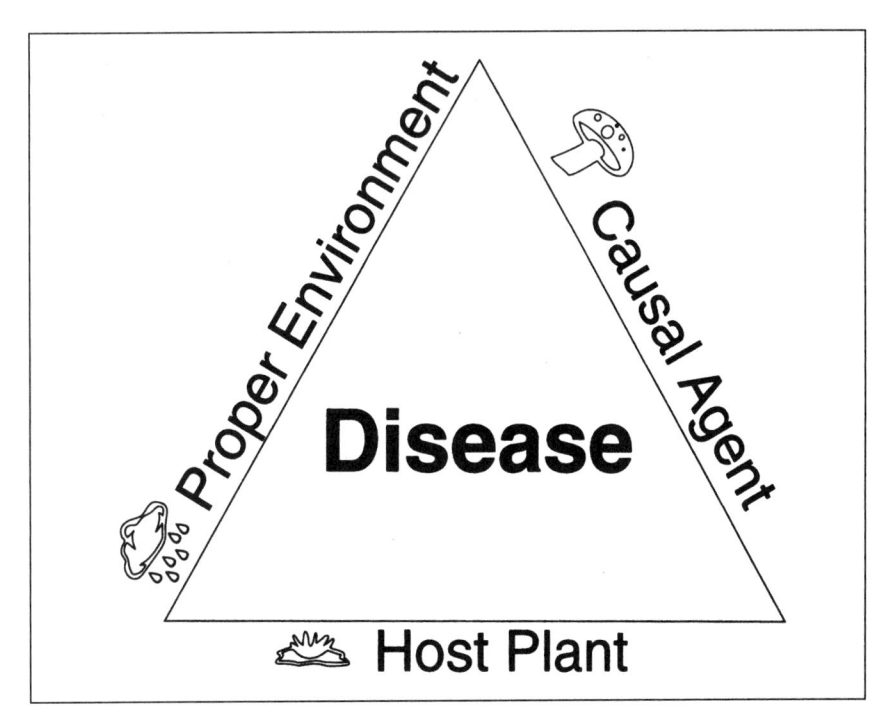

FIGURE 9-6 The disease triangle

Like any other organism, disease pathogens are active within a particular range of environmental conditions. If the environmental conditions are not present, a disease is not likely to occur. Modification of the environment is one means of disease prevention or control. For example, providing good air circulation and soil water drainage on golf greens decreases the incidence of Phythium Blight.

As with other pest problems, the turf manager should be familiar with the different turf diseases, the conditions under which diseases will likely occur, and various identifying characteristics of disease problems. This knowledge will enable the turf manager to implement appropriate controls.

ACTIVITIES

Weed Scavenger Hunt

Equipment.

- turf weed reference guide or turf weed ID chart
- collection bag
- hand trowel

Procedure.

In designated maintained turf areas, identify and collect the following:

- two grassy weeds
- five broadleaf weeds

The weeds must be from a maintained turf area that is subject to regular mowing and other cultural practices. Tall, uncut weeds from ditches or alongside buildings should not be used. Use various field references to properly identify the weeds. While collecting, identify the weeds before digging and placing them in collection bags. When you have collected the weeds, return to your instructor to receive proper credit for the activity.

Turf Weed Collection

Using the common turfgrass weeds table (Table 9-1), develop a collection of twenty-five weeds. Use either the pressed-specimen method or photo journal method as assigned by your instructor. At the time of collection, record pertinent data, including botanical name, common name, date, location (city, state, county), collector name, locality (lawn, sports field, golf course) and habitat (full sun, shade, damp soils).

Pressed Specimens. Collect and preserve weeds found in maintained turf areas. Mount the weeds on three-hole punched, 8.5 inch × 11 inch light card-stock paper. Label the collection sheets using the information recorded while collecting the weeds. Place the completed collection in your three-ring binder for later reference.

Procedure.

1. Select three to five single weed plants (not clumps) found in a *maintained* turf area. Plants may be removed from the soil using a variety of digging implements including a knife or trowel. At the time of collection, record pertinent data, including name of plant, date of collection, location (city, state, county), collector name, locality (lawn, sports field, golf course) and habitat (full sun, shade, damp soils).

2. Place the collected specimens within single folds of newspaper. Arrange the leaves of the specimens so that both the front and back sides of the leaves can be observed. All soil must be removed from the roots. Washing may be necessary to remove soil.
3. Place the specimens, now within the newspaper, between two pieces of blotter paper or corrugated cardboard, and then place the whole arrangement under heavy weights, such as several textbooks. (Do not use the turf textbook.)
4. Drying should occur within three to eight days. Check daily to ensure that mold and mildew do not start growing on your specimens. Change the blotter paper or cardboard as necessary for drying. Do not try the following methods: hair dryers, heating vents, conventional ovens, or microwave ovens. None of these methods works effectively.
5. When dry, select one or two single plants for mounting. Showing front and back arrangements, mount the specimens in the spaces provided on the herbarium sheet (Figure 3-11), using paste, muscilage, linen tape, white glue, or transparent adhesive film. Do not use common cellophane tape, transparent tape, rubber cement, epoxy or plastic adhesives (super glues).
6. Label each card in the lower right corner with the appropriate information. Type or print neatly with black ink *before* mounting specimens.
7. Turn in the sheets for evaluation on the date provided by your instructor.

Photo Journal. Using color-print film, photograph weeds found in maintained turf areas. Mount two photographs on one sheet of light-card-stock paper. Label each photograph using the information recorded while photographing the weeds. Place the completed collection in your three-ring binder for later reference.

Turf Insect Collection

With the assistance of your instructor, identify, collect, and mount the various insects that cause problems in a turf. Photograph the damage caused by the insect pest for inclusion in your collection.

Turf Disease Study

View photographs of the signs and symptoms of different turf diseases. Study prepared microscopic slides of the various disease pathogens. If conditions are favorable, study actual disease problems in turf areas of the campus and the surrounding community. Photograph the disease problem for inclusion in your collection.

EXERCISES

Turf Pest Records

Develop a file of various turf pest problems using a form you developed in earlier pest-collection activities. Include the weed collections and photographs of insect and disease problems from the different activities.

Using Paper. Using duplicated copies of your form, develop a file of the pest problems to keep in your three-ring binder. Using the information on each pest problem, create a binomial key and index to the different pest problems.

Using a Computer. Using the sample form as a template for the record layout, develop a database of the pest problems. Set up a query to identify the pest problem based on characteristics, signs, and symptoms.

On-Line Companion. U09E01.WDB—Pest problem disease

UNIT 10

Turf Pest Control

Even under the best cultural care, there is a chance that pest problems will occur. A turf is a dynamic ecological system. The less complex the system, such as a turf with one to a few different grasses, the more likely that another population, such as an insect pest, will potentially overwhelm the system. It is important for the turf manager to maintain a proper balance in the turfgrass **ecosystem**. The turf manager should strive to maintain a healthy turf capable of withstanding pest problems. Continual monitoring for pest problems allows the turf manager to apply the appropriate controls while limiting impact on other parts of the system.

For many years, the first line of defense against pests was the use of chemicals. Given the growing concern regarding the potentially detrimental impact of chemicals on the environment, today's turf manager needs to utilize a wide range of control measures, including nonchemical and biological as well as chemical measures.

PEST CONTROL THROUGH PROPER CULTURE

In many turf areas suffering from a pest problem, the initial cause of the problem often has its roots in poor or improper cultural practices. Due to these poor practices, the turf is under a stress that hinders the grass from withstanding invading pests. A turf manager can prevent or control many pest problems by applying proper cultural practices.

For example, to control annual crabgrass growth, an Extension agent should advise homeowners to raise the cutting height on their mowers. By raising the cutting height, the germinating crabgrass seed cannot grow because of shading from the taller, denser turfgrass.

Many pest problems can be attributed to improper cultural practices. Improper mowing heights and improper mowing frequency are just two examples of poor cultural practices that can lead to increased weed problems. When establishing methods of pest control, the turf manager should thoroughly examine the current cultural practices and adjust any improper practices to give the turf an advantage over pest organisms.

MONITORING FOR PEST PROBLEMS

An important aspect of pest control is monitoring and anticipating any pest problems. Pests are much easier to control in the early stages of an attack. It is, therefore, important that both the turf manager and the turf maintenance crew be able to monitor for and recognize any developing pest problems.

With regard to agricultural crops, there is a level of pest attack known as the "economic threshold." The goal is not 100 percent control of pest. Rather, the grower continually monitors the fields. When the pest population is to a level as to cause an economic loss, the grower implements pest control measures.

With turf, the threshold varies depending on the purpose of a turf. For a home lawn, the threshold is often aesthetic in nature; the homeowner will apply controls when the turf "starts to look bad." Further, the range of aesthetic thresholds can be very board among homeowners.

With sports turf, the threshold relates both to function and economics. This is very apparent when considering a golf course turf. A pest attack on a golf green can disrupt the playing surface, thereby reducing the putting quality of the green. If golfers have problems putting on a pest-damaged green, they will likely go to another golf course. The loss of paying golfers would, in turn, have an economic impact on the pest-ridden golf course.

Although the term *monitoring* often calls to mind the image of walking around a turf and searching for the pest organisms, this is only one form of monitoring. Environmental monitoring is another way of forecasting pest problems.

In recent years, turf researchers have found that many turf diseases and some insect pests are most likely to occur when a certain combination of environmental conditions fits a predictor-model equation. For example, a combination of temperature and length of leaf surface wetness may indicate a high chance that Anthracnose disease will occur within a few days. Only when the model indicates a high probability of pest attack does the turf manager need to implement the appropriate controls, in particular, chemical controls. As long as the predicator model gives a low chance of pest attack, the turf manager can keep any pesticides in storage.

NONCHEMICAL PEST CONTROL

In addition to the use of chemical pesticides, there are numerous methods the turf manager can employ to control pests. As previously mentioned, proper cultural practices are major factors in the prevention and control of many turf pest problems.

Another method of pest control takes place during establishment or **renovation** of a turf. Use of pest-resistant turfgrass varieties or species can aid in reducing the potential for pest problems. Researchers have found that endophyte-infected ryegrass and tall fescue prevent and control the incidence of sod webworm and other insect pests. Breeders are selecting and growing many newer turfgrass varieties that are resistant to several turf diseases.

For weed problems, physically pulling out the weed has been a traditional method of control. For many broadleaf weeds, this is a simple method that results in very little impact on the environment. The labor involved in pulling weeds restricts this practice to small turf areas, however.

BIOLOGICAL PEST CONTROL

Biological pest control involves the use of other organisms to control the pest organisms. The effectiveness of biological control varies depending on the pest problem and, of course, whether a biological control agent is available for use against the pest.

The greatest success to date has been in the control of some insect pests. The use of endophyte-enhanced turfgrass is considered a type of biological control. The endophyte inside the turfgrass produces a compound that is toxic to insects that feed on the turfgrass plant. A common biocontrol method is use of milky spore disease, a variety of *Bacillus sp.* bacteria, to control grubs.

Milky spore bacteria invade certain root-feeding grubs. When inside the grub's body, the bacteria produce a toxin that destroys the grub. Laboratories can produce large quantities of the bacteria for application to a turf.

Another method of biocontrol is the use of parasitic nematodes. Nematodes attack specific insect pests for a source of food. In order to maintain any degree of control when using nematodes, however, care must be taken in handling and application. Results with the use of nematodes for turf pest control have been variable to date.

Biological weed control has been insufficient to date. Some researchers have done work using bacteria to control annual bluegrass, but with very little success. Many of the pests of weeds that have potential for control are also pests of turfgrass.

Due to the nature of the disease, it has been difficult to find and develop any biological control agents to combat disease-causing organisms. An area that shows promise is the use of compost to increase a diverse population of soil organisms in a turf. This appears to have an antagonistic effect against disease organisms living in the soil.

CHEMICAL PEST CONTROL

Chemical pest control involves the use of chemicals, called pesticides, that are toxic to pests. Depending on the pest and the specific action of the chemical, pesticides can be specific to a single pest or very broad in controlling a large number of organisms—including the desirable turfgrass plants.

Insecticides

Insecticides are pesticides for the control of insects. There are several mechanisms of action among the various insecticides. Some insecticides are nerve toxins, which disrupt the nervous system of the insect. Other insecticides are stomach poisons, which destroy the lining of the insect's stomach.

A new group of insecticides that is being used on an increasing basis are *insect growth regulators* or *IGRs*. Insect growth regulators regulate or stop the growth of an insect to prevent it from completing its life cycle. Precor (methoprene) is an IGR that is used for the control of fleas in carpets and lawns.

Fungicides, Nematicides, and Antibiotics

Because fungi cause most turf diseases, fungicides are the most commonly used disease-control pesticides. Nematicides are for control of disease-causing nematodes, while antibiotics are for control of the few bacteria and viruses that may attack a turf.

The turf manager uses fungicides in either preventive or curative applications. Preventive applications are performed prior to a disease attack, when the turf manager anticipates that conditions will be favorable for an attack. A turf manager will apply curative applications when the disease is present on a turf.

The time of application depends on the use of the turf. For golf course greens, where a high-quality putting surface is necessary, superintendents often apply fungicides on a preventive basis regardless of the presence or absence of disease-causing factors. For lawns and utility turf, the turf manager can apply fungicides on a curative basis and at the same time apply appropriate cultural practices to encourage recovery of the turf.

Fungicides are classified as either contact or **systemic**. Contact fungicides are applied to the surface of the turfgrass leaves to create a toxic barrier against the disease organism. Systemic fungicides are absorbed by the turfgrass plant to create a situation that is toxic to invading disease organisms.

Contact fungicides often give a broader spectrum of control over several different turf disease organisms than do the systemic fungicides. Application of contact fungicides must be more frequent because these fungicides can break down from sunlight or be knocked off leaf blades by mechanical action.

Systemic fungicides persist for a longer time because the fungicide compounds are inside the plant. Systemic fungicides often persist four to six times longer than similar contact fungicides. Efficiency decreases over time primarily due to mowing. It should be noted that resistance of some disease organisms, such as Dollar Spot, to certain systemic fungicides is an ongoing problem.

Herbicides

Herbicides are compounds that are toxic to plants. **Selective herbicides** control only certain weeds and have little or no impact on surrounding desirable plants. Herbicides used to control broadleaf weeds in turf are examples of selective herbicides. **Nonselective herbicides** control all plants that come in contact with the herbicide. Glyphosate (Roundup®, Kleenup®) is an example of a herbicide that will control most plants found in a turf, including the desirable turfgrasses.

The turf manager should identify the target weeds in order to determine time and method of application. For some weeds, such as annual grasses like crabgrass, pre-emergent herbicide is a practical chemical control. For other weeds, particularly the broadleaf weeds, **post-emergent herbicides** provide effective control.

Pre-emergent herbicides work on the principle that germinating plants are more susceptible than mature plants to low levels of the herbicide. The turf manager applies pre-emergent herbicides to a turf in order to form a toxic barrier to germinating seeds. When applied to a mature turf stand at the proper rates, pre-emergent herbicides have no effect on the turfgrass. Timing of application is often critical, particularly with crabgrass. After the seed has germinated and put out one or two leaf sets, the pre-emergent herbicide is no longer effective.

Post-emergent herbicides are usually applied to actively growing weeds. Many post-emergent herbicides are also known as plant growth regulators, because they alter or disrupt the growth processes in the plant. Many selective post-emergent herbicides work because of the physiological difference between turfgrass and other weeds, usually broadleaf plants. Many post-emergent herbicides cannot, however, control grassy weeds because of the physiological similarities between turfgrass and these weeds.

Formulations

Pesticides are available in a variety of **formulations** ranging from dry granular products to liquid products. The formulation a turf manager chooses depends on several factors, not the least of which is what formulations are offered by the manufacturers. Some pesticides come only in liquid form, while others are available in several different dry or liquid formulations.

Granular-Applied Pesticides. Dry, granular formulations are best when complete uniform coverage of the turfgrass plant is not critical. Such is the case when applying pesticides that require watering into a turf, such as insecticides for control of root-feeding insects. Most granular pesticides are similar to fertilizers in that the turf manager can use the same spreader equipment to apply both types of material.

Many fertilizer companies sell products that are a combination of fertilizer materials and granular pesticides. Fertilizer, grub control insecticides, and pre-emergent herbicides often require application at the same time. These combination products are useful at certain times of the season and the single application reduces required labor.

Liquid-Applied Pesticides. Liquid-applied pesticides provide more uniform coverage of the pesticide over a turf than do granular pesticides. Although requiring more equipment, sprayed pesticides allow for exact placement of the material, resulting in more effective control of some turf pests.

There are several different formulations of liquid-applied pesticides. With regard to actual pest control, there is not a drastic difference between the different liquid-applied pesticide formulations. Those differences that are important to the turf manager involve such factors as the calculations required to determine the amount of active ingredient to mix, and the impact that the formulations have on sprayer equipment (such as wettable powders causing spray nozzle wear).

Many granular pesticides can also be applied as liquids, the only difference being that the pesticide is already dissolved in water rather than depending on rainfall or irrigation in order to dissolve.

ACTIVITIES

Pest Monitoring

With assistance from your instructor, monitor for various pest problems in a turf area on campus. Use a variety of monitoring techniques including baits, traps, and soil drenches. Keep a record of any identified pest. Note when these pests approach a pre-defined threshold.

Nonchemical Pest Control

In an area known to have a bluegrass billbug problem, **overseed** various plots with an endophyte-enhanced turfgrass species. (This could be perennial ryegrass, tall fescue, or a combination of both species.)
   Over the growing season, monitor for pest attack and evaluate the turf areas for degree of pest attack.

Pest Control Research

Attend a turf research field day, visit a nearby turf research facility, or attend a turf conference to learn about current studies in biological and chemical pest control.

EXERCISES

Pesticide Inventory

Develop a file of all pesticides used for pest control on the campus turf areas. Include all information required to meet various federal, state, or local regulations. As materials are used on campus turf areas, adjust the inventory as necessary.

Using Paper. Develop a file using a form similar to the sample pesticide inventory form (Figure 10-1). Maintain the file in your three-ring binder. Organize the records so that you can quickly find any given pesticide when updating the inventory.

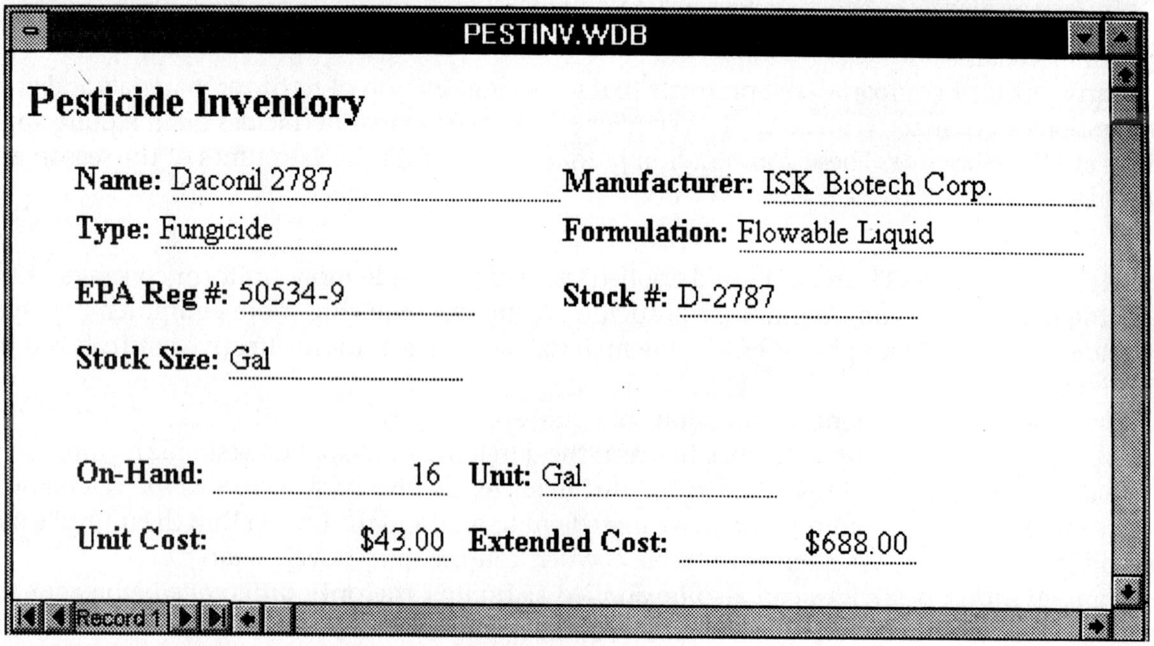

```
PESTINV.WDB
```

Pesticide Inventory

Name: Daconil 2787   Manufacturer: ISK Biotech Corp.
Type: Fungicide   Formulation: Flowable Liquid
EPA Reg #: 50534-9   Stock #: D-2787
Stock Size: Gal

On-Hand: 16   Unit: Gal.
Unit Cost: $43.00   Extended Cost: $688.00

Record 1

FIGURE 10-1 Sample pesticide inventory form

Using a Computer. Develop a data file for maintaining the pesticide inventory. Create a record layout similar to the sample pesticide inventory form (Figure 10-1). Include a query form to allow you to quickly find any given pesticide when adjusting the inventory.

On-Line Companion. U10E01.WDB—Pesticide inventory database

Pesticide Calculations

Develop a means of calculating the amount of pesticides to be applied to the various campus turf areas. Use the information from previous unit activities and exercises, such as campus site area measurements, along with the information contained in the pesticide inventory file.

Using Paper. Develop a form for performing pesticide calculations. Use the information in the pesticide inventory as a reference source to complete the calculations.

Using a Computer. Develop a spreadsheet to perform the pesticide calculations. If the software permits data file lookups, use the information contained in the pesticide inventory.

On-Line Companion. U10E02.WKS—Pesticide calculations spreadsheet

Pesticide Application Record File

Develop and maintain a file to keep track of all pesticide applications performed on campus turf areas. Include all information required to meet various federal, state, or local regulations. On a monthly basis, create a report summarizing all the pesticide applications performed on campus.

Using Paper. Develop a file using a form similar to the sample pesticide application record (Figure 10-2). Set up an additional form for the monthly report. Keep all records and reports in your three-ring binder.

Using a Computer. Develop a data file with a record layout similar to the sample pesticide application record (Figure 10-2). Set up the database to generate a monthly report of pesticide usage.

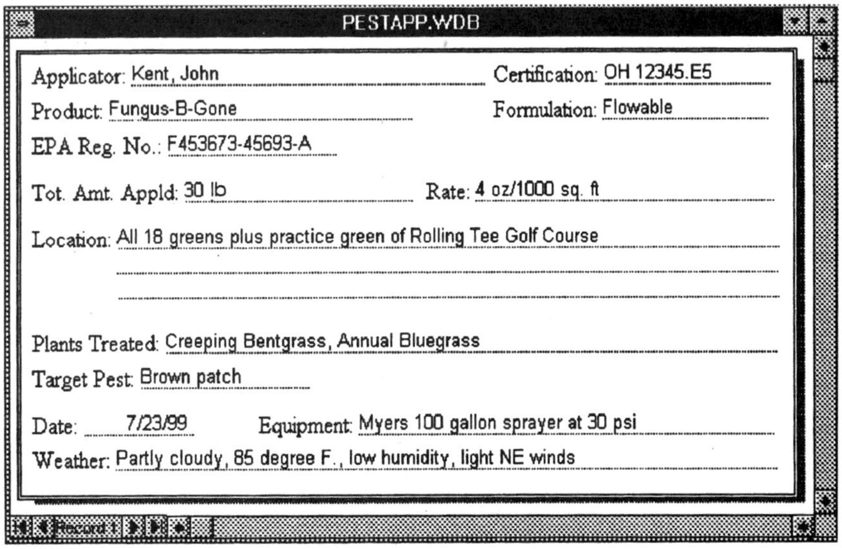

PESTAPP.WDB

Applicator: Kent, John Certification: OH 12345.E5

Product: Fungus-B-Gone Formulation: Flowable

EPA Reg. No.: F453673-45693-A

Tot. Amt. Appld: 30 lb Rate: 4 oz/1000 sq. ft

Location: All 18 greens plus practice green of Rolling Tee Golf Course

Plants Treated: Creeping Bentgrass, Annual Bluegrass

Target Pest: Brown patch

Date: 7/23/99 Equipment: Myers 100 gallon sprayer at 30 psi

Weather: Partly cloudy, 85 degree F., low humidity, light NE winds

FIGURE 10-2 Sample pesticide application record

On-Line Companion. U10E03.WDB—Pesticide application database

Pesticide Management System

Develop a comprehensive pesticide management system to include inventory, calculations, applications, and any other information required by government regulations, such as materials safety data sheets (MSDS). Include instructions as to the proper use of the system.

Using Paper. Develop an organized record manual complete with pre-printed, indexed forms with necessary labeling, and instructions for using the manual.

Using a Computer. Develop the system using a relational database management system (DBMS). Include all the necessary entry forms, display forms, and data entry forms for a basic "turnkey" system.

UNIT 11

Turf Materials Application

To obtain a desired result, the turf manager must uniformly apply many materials to a turf area. Misapplication of a material can be detrimental to a turf; and bringing the turf back to a desired quality level often requires additional expenditure. With regard to pesticides, misapplication could lead to environmental and safety problems that could be of major financial consequence to the turf manager.

There are several methods for applying materials to a turf. These methods usually fit into the major categories of dry applications and liquid applications. Turf specialists often will refer to the application of dry fertilizers and pesticides as granular applications. Turf managers often will use the same equipment to apply granular materials as they do to plant turfgrass seed during establishment and, occasionally, to uniformly spread **topdressing** materials such as sand and screened composts.

Regardless of the method of application, it is important to calibrate the application equipment to apply the material at the desired rate. Even if a homeowner or turf manager buys the correct total amount of material for a turf area, lack of proper equipment calibration can lead to an uneven distribution of material and, thus, to poor results. Further, different materials, corrosion, the applicator, and other factors can alter the original calibrated setting of the application equipment.

FERTILIZER AND PESTICIDE APPLICATION METHODS

There are two major methods of applying fertilizers and pesticides to a turf area. The most common is dry application using granular fertilizers; the next most common is liquid application. The method used will depend on several factors including the size of the turf area, equipment, labor, purpose of application, and the desired effects on a turfgrass.

Granular Application Equipment

The two major types of spreaders, *drop spreaders* and *rotary spreaders* (Figure 11-1), are the most commonly used equipment a turf manager will use in the application of dry granular materials (as well as of other dry materials such as topdressing and seed). The size of the area and the material to be applied influence a turf manager's selection of the type of spreader.

Drop Spreaders. A drop spreader, or gravity spreader, consists of a steel or plastic hopper. Along the bottom length of the hopper is a series of holes. An adjustable plate or gate covers the holes and serves as a means of controlling the flow of material out of the hopper.

Drop spreaders provide a fairly uniform application of material over a turf surface. There is, however, a limit to the width of the spreader and, therefore, the application swath. Most drop spreaders for turf are between two feet and three feet in width, although some spreaders are available up to six feet in width. Because of their small size and application swath, drop spreaders are useful for small turf areas that cannot easily be covered by other application means.

FIGURE 11-1 Two major granular spreaders for turf use

Rotary Spreaders. A rotary or centrifugal spreader consists of a hopper with a large single hole. Beneath the opening is a spinning impeller. As the granular material drops onto the spinning impeller, the granules fly outward by centrifugal force.

Even though the hopper is small, this rotary action allows for a wide application swath ranging from five feet to fifty feet depending on the size of the spreader. As with the drop spreader, an adjustable plate at the base of the hopper controls the flow of material onto the impeller.

A variation of the rotary spreader is the pendulum spreader. Instead of a spinning impeller, a swinging tube attaches to the base of the hopper. As the tube swings in a pendulum-type action, material flies out by centrifugal force in a pattern very similar to that of a rotary spreader. Operating the pendulum spreader requires a power take-off, which restricts the spreader to mounting on power machinery.

Because of the wide application swath, rotary spreaders can cover large areas in a relatively short time. Uniformity of application may not be as good as with drop spreaders, however. One reason for nonuniform coverage is the wind causing particle "floating." Also, particles of varying density, as are found in mixed or bulk-blended fertilizers, tend to drop out at different distances from the spreader, which results in a banding effect.

Specialty Materials Application Equipment. Golf course managers and other large-area turf managers will use special application equipment for topdressing and seed sowing. Topdressers are machines with a conveyor belt-type hopper bottom and a spinning brush on the end. Topdressers permit uniform application of the heavy and bulky topdressing material that most gravity or rotary spreaders cannot spread.

Although drop or rotary spreaders can be used for seed sowing, seeders, which usually are modified drop spreaders, are often used for sowing seed into a turf area. These machines can perform a variety of functions, including soil preparation, seed placement, and seed coverage, all in one pass.

Liquid Applications

Liquid applications are useful in some turf situations. Unlike granular materials, liquid fertilizer and many pesticides, do not require rainfall, irrigation, or soil water to dissolve the material. Liquid pesticide applications allow for placement of the material directly on the target plant. With weeds, these applications permit direct chemical control of the plant; with disease and insects, they serve as a means of protecting or curing the plant of the pest problem.

Foliar Spray. Foliar application involves applying the material with small amounts of carrier, usually water, so that the material coats the leaves. Foliar application requires a spray apparatus consisting of one or more nozzles and a tank, piping, and pressure-pumping system.

Soil Drench. A soil drench involves mixing water-soluble material in a large enough amount of water to carry the material into the soil. The goal is to place the material in the upper soil layer beneath the turfgrass plants. Turf managers apply pre-emergent herbicides, liquid fertilizers, and pesticides as a soil drench for controlling soil insect pests or soil-borne fungal diseases.

Irrigation Injection. Irrigation injection is a variation of soil drenching wherein the material is injected into the irrigation system. For fertilization, the term is *fertigation* or *fertilizer injected irrigation*. This method allows the turf manager to use the irrigation system as an alternative to sprayers and spray tanks. Turf researchers and turf managers have had a varying degree of success using irrigation systems to apply fertilizers, insecticides, and fungicides to turf.

Obtaining uniform application requires extensive work in proper irrigation system design and installation. There are also several potential environmental and health hazards that must be controlled when using large scale irrigation injection. Fertigation and pesticide-injected irrigation are, therefore, best left to other areas of horticulture and agronomy.

CALIBRATION

Calibration is the process of adjusting the output rate of the spreader or sprayer to equal the desired rate of material application for a turf area. Calibration is very critical to obtain proper and uniform application of the material. With regard to pesticides, several governmental rules and regulations require very accurate application of the material. The turf manager must use only the amount required to control the pest problem. Damage to desired plants often results from the failure to accurately calibrate a sprayer for herbicide application.

Because each material has different properties, such as analysis, particle density, particle size, or flow characteristics, spreader and sprayer calibration setting will be different for each material. A homeowner using a sprayer or spreader calibrated for one material, such as urea fertilizer, may not obtain the same coverage when applying other materials, such as combination products. For every change of product, the spreader or sprayer should be re-calibrated to obtain accurate application of material. Even fertilizers with the same analysis but different carriers will require separate spreader calibrations because of potential differences in particle size and density.

Each time a spreader or sprayer is calibrated for a particular material, the information should be recorded in a record book for later reference. In some localities, pesticide application record-keeping regulations may require the turf manager to keep calibration records.

In addition, due to the corrosive or abrasive effects of many of the materials, it is important to check calibration of equipment on an annual basis. And if a mechanic performs any maintenance on the equipment, such as overhauling a pump or painting the spreader hopper, the turf manager should recheck the equipment for proper calibration.

SPREADER CALIBRATION

Regardless of the material, spreader calibration is the same for any type of granular product. The goal is to have material flowing out of the hopper so that the amount of material applied to the surface will match the desired rate and coverage of material.

Drop Spreader Calibration

In-Place Calibration.

1. Obtain a sheet of plastic or heavy paper larger than the spreader, and place it on a clean, level surface.
2. Using blocks or other means of support, prop the spreader frame above the sheet so that both wheels turn freely.
3. Determine the circumference of the wheel (*circumference* = $2\pi r$).
4. Calculate the number of complete rotations to travel a linear distance such as 100 feet (*rotations = distance / circumference of wheel*).
5. Mark the wheel with chalk and fill the hopper with the material.
6. Open the hopper and turn the wheels the predetermined number of rotations. If there is a split agitator in the hopper, turn both wheels in the same direction.
7. Weigh the fertilizer collected on the sheet.
8. The area covered is *distance × spreader width*
9. The application rate based on a standard weight per unit area (pounds per 1,000 square feet or kilograms per 100 square meters) is determined using the formula:

    rate = (amount collected / area covered) × unit area

    or, in English measurements of pounds per 1,000 square feet:

    rate = (amount collected / area covered) × 1,000 square feet

10. Adjust the rate setting and repeat the process until the spreader is calibrated to the desired rate.
11. Record the material, rates, and settings in a record book for later reference.

Catch Pan Method.

1. Using cardboard, sheet metal, plastic, or PVC pipe make a catch pan to cover the entire length of the hopper.
2. Attach the pan to the bottom of the hopper using hooked elastic cords (bungee cords). Make sure the spreader controls continue to operate properly.
3. Operate the spreader over a measured area, avoiding any overlap.
4. Weigh the material collected in the catch pan.
5. The application rate based on a standard weight per unit area (pounds per 1,000 square feet or kilograms per 100 square meters) is determined using the formula:

    rate = (amount collected / measured area) × unit area

    or, in English measurements of pounds per 1,000 square feet:

    rate = (amount collected / measured area) × 1,000 square feet

6. Adjust the rate setting and repeat the process until the spreader is calibrated to the desired rate.
7. Record the material, rates, and settings in a record book for later reference.

Rotary Spreaders

Remainder Method.

The remainder method works well for small spreaders that can be easily emptied.

1. Select a setting for the spreader and make appropriate adjustments. If available, use the manufacturer's recommendations as starting points.
2. Fill the hopper with a known weight of material, such as a 50-pound bag.
3. Apply material to a specific, measured area (usually a 1,000 square-foot) area.
4. Weigh the remainder of the material in the spreader. The difference between the starting weight and final weight is the amount of material applied with the spreader.
5. The application rate based on a standard weight per unit area (pounds per 1,000 square feet or kilograms per 100 square meters) is determined using the formula:

$$\textbf{rate} = (\text{amount applied} / \text{area covered}) \times \text{unit area}$$

or, in English measurements of pounds per 1,000 square feet:

$$\textbf{rate} = (\text{amount applied} / \text{area covered}) \times 1{,}000 \text{ square feet}$$

6. Adjust the rate setting and repeat the process until the spreader is calibrated to the desired rate.
7. Record the material, rates, and settings in a record book for later reference.

Replacement Method.

The replacement method is for large, tractor-drawn or mounted rotary spreaders that cannot be easily emptied for weighing purposes.

1. Select a setting for the spreader and make appropriate adjustments. If available, use the manufacturer's recommendations as starting points.
2. Mark the spreader hopper at a given level and fill with material to that mark.
3. Apply material to a specific, measured area (usually a 1,000-square-foot area).
4. Weigh the amount of material needed to refill the spreader.
5. The application rate based on a standard weight per unit area (pounds per 1,000 square feet or kilograms per 100 square meters) is determined using the formula:

$$\textbf{rate} = (\text{amount collected} / \text{area covered}) \times \text{unit area}$$

or, in English measurements of pounds per 1,000 square feet:

$$\textbf{rate} = (\text{amount collected} / \text{area covered}) \times 1{,}000 \text{ square feet}$$

6. Adjust the rate setting and repeat the process until the spreader is calibrated to the desired rate.
7. Record the material, rates, and settings in a record book for later reference.

Clean Sweep Method.

The clean sweep method is similar to drop spreader calibration.

1. Select a setting for the spreader and make appropriate adjustments. If available, use the manufacturer's recommendations as starting points.

2. Fill the hopper with material.
3. Operate the spreader over a clean, swept area such as a concrete parking lot.
4. Sweep up and collect all the material in a measured area within the overall spreader area.
5. Weigh the material.
6. The application rate based on a standard weight per unit area (pounds per 1,000 square feet or kilograms per 100 square meters) is determined using the formula:

$$\textbf{rate} = (\text{amount collected} / \text{area covered}) \times \text{unit area}$$

or, in English measurements of pounds per 1,000 square feet:

$$\textbf{rate} = (\text{amount collected} / \text{area covered}) \times 1{,}000 \text{ square feet}$$

6. Adjust the rate setting and repeat the process until the spreader is calibrated to the desired rate.
7. Record the material, rates, and settings in a record book for later reference.

Catch Sheet Method.

The catch sheet method is a variation of the clean sweep method.

1. Select a setting for the spreader and make appropriate adjustments. If available, use the manufacturer's recommendations as starting points.
2. Fill the hopper with material.
3. Operate the spreader over a sheet of plastic, cloth, or paper, such as a painter's drop cloth.
4. Measure the area of the sheet.
5. Collect and weigh the material on the sheet.
6. The application rate based on a standard weight per unit area (pounds per 1,000 square feet or kilograms per 100 square meters) is determined using the formula:

$$\textbf{rate} = (\text{amount collected} / \text{area of collection sheet}) \times \text{unit area}$$

or, in English measurements of pounds per 1,000 square feet:

$$\textbf{rate} = (\text{amount collected} / \text{area of collection sheet}) \times 1{,}000 \text{ square feet}$$

7. Adjust the rate setting and repeat the process until the spreader is calibrated to the desired rate.
8. Record the material, rates, and settings in a record book for later reference.

SPRAYER CALIBRATION

There are several methods for calibrating a sprayer. Some are very similar to methods for calibrating a spreader. Several factors impact calibration including supply pressure, speed, and nozzle size.

In all methods of sprayer calibration, the material should be used instead of water because the density of the material solution may result in a flow rate that is different from the flow rate for water. To prevent potential injury to the turf test area, water can be used for initial settings of the equipment and test runs. The actual material should then be used to fine-tune the calibration setting.

When operating spray equipment, one should have the machinery or the applicator for hand sprayers in motion (a "running start") before opening the sprayer valve. During calibration, use of a tracer dye may aid in seeing and measuring application areas.

Boom Sprayers

Collection Jar Method.

1. Select a speed and pressure setting.
2. With the sprayer in place, collect liquid from the nozzles for one minute, and calculate an average volume per nozzle output (in gallons or liters).
3. At the selected speed, operate over an area for one minute using the same pressure setting. Measure the traveled distance.
4. The test application area is the *boom width × distance traveled.*
5. Determine the output rate using the formula:

    rate = ([average volume per nozzle × number of nozzles] / test application area) × unit area

    If working in gallons per acre, the unit area must be converted to a uniform dimension: (that is, the unit area will be 43,560 square feet if the test application area is in square feet).

6. Adjust various factors such as speed, pump pressure, or nozzle size, and repeat the process until the sprayer applies the desired rate.
7. Record the material, rates, nozzle sizes, and settings in a record book for later reference.

Replacement Method.

1. Select a speed and pressure setting.
2. Fill the tank to a marked capacity.
3. At the selected speed, operate the sprayer over a measured test application area.
4. If the tank has an accurate volume gauge, note the new volume and determine the volume of applied spray material (original volume mark − new volume mark). If the tank does not have a volume gauge or level, refill the tank to the original marked capacity and note the replacement volume.
5. Determine the output rate using the formula:

    rate = (volume of applied material / test application area) × unit area

    If working in gallons per acre, the unit area must be converted to a uniform dimension (that is, the unit area will be 43,560 square feet if the test application area is in square feet).

6. Adjust various factors such as speed, pump pressure, or nozzle size, and repeat the process until the sprayer applies the desired rate.
7. Record the material, rates, nozzle sizes, and settings in a record book for later reference.

ACTIVITIES

Spreader Calibration

Using one or more of the methods listed in the unit, calibrate a drop spreader and a rotary spreader to apply a desired rate of a granular material (fertilizer, pesticide, topdressing). Your instructor will provide the basic rate of active ingredient or nutrient. Use the calculation methods in previous units to convert from rate of nutrient or active ingredient to rate of material.

Sprayer Calibration

Using one or more of the methods listed in the unit, calibrate a hand-held spray gun and a boom sprayer to apply a desired rate of a liquid material (fertilizer, herbicide, pesticide). Your instructor will provide the basic rate of active ingredient or nutrient. Use the calculation methods in previous units to convert from rate of nutrient or **active ingredient** to rate of material.

EXERCISES

Calibration Records

Develop a form for maintaining calibration records of the equipment for the various materials used in the care of the campus turf areas.

Using Paper. Develop a form using the layout in the sample calibration record (Figure 11-2). Add any additional fields as necessary. Enter the information collected in the spreader and sprayer calibration activities. Use one copy of the form per calibration. Keep the completed forms in your three-ring binder.

Using a Computer. Develop a database file using the form layout in the sample calibration record (Figure 11-2). Add any additional fields as necessary. As an advanced exercise, develop a one-to-many relational database with application equipment information kept in one data file and linked to a data file with entries of the different materials, rates, and settings.

On-Line Companion. U11E01.WDB—Materials calibration database

Spreader Calibration Record					
Manufacturer:		Model:		Type:	
Application Width:		Application Speed:		PTO Speed:	
Date		Material	Analysis	Rate	Setting

FIGURE 11-2 Sample spreader calibration record

UNIT 12

Turf Cultivation and Renovation

Unlike a farmer, the turf manager cannot use a plow to cultivate established turf. To correct underlying soil problems in a turf, the turf manager must use specialized cultivation equipment in place of plows, tillers, and discs. The equipment used by the turf managers includes core aerifiers, spikers, slicers, and similar equipment.

Thatch is a layer of plant tissue that develops between the green grass and the soil surface of most creeping-type grasses. In many turf situations, the thatch layer becomes thick enough to cause problems. Many of the methods for controlling thatch involve cultivation practices.

Over time, a turf may start to thin out for various reasons. Instead of completely removing the old turf, the turf manager can renovate a turf by using cultivation equipment to place new seed within an existing turf stand.

CULTIVATION

There are many reasons to cultivate a turf. Most of the reasons involve improving soil conditions beneath the turf layer. In many turf situations, particularly on sports fields, golf courses, and other high traffic areas, the soil becomes compacted. This results in reduced water infiltration and reduced soil aeration, which threaten root survival. Compacted soils also create mechanical barriers to root growth.

In addition, there are several cultivation-like practices that enhance turfgrass growth even though the soil is not in a compacted state. Turf managers, particularly golf course superintendents, use cultivation machines to slice lateral stems as a way to encourage dense shoot growth. Managers use other cultivation practices to improve the movement and placement of chemicals in the soil, such as pesticides for control of root-feeding insects.

Core Aeration

Core aeration (also known as **coring, core cultivation**, and core aerification) is a method whereby machines extract small soil cores from a turf (Figure 12-1). These cores range from 1/4 inch (0.64 cm) to 3/4 inch (1.9 cm) in diameter and are usually 2 inches (5.1 cm) to 4 inches (10.2 cm) in length. Some deep-tine core aerators can extract cores upwards of 12 inches (30.5 cm) in length.

Turf managers use core aeration for several different purposes. The prime use is to relieve compaction and improve soil structure. Extracting core from a soil causes the surrounding soil to collapse into the remaining holes, resulting in a loosening effect. Turf roots tend to grow through the holes, allowing for deeper root penetration. The improved root growth, in turn, enhances soil structure development.

By repeatedly removing cores from a turf area over a long time and replacing the soil with a coarser-textured material through topdressing, a turf manager can gradually shift the underlying surface soil texture. Golf course superintendents often perform this practice on old, native-soil "push up" golf greens that cannot be taken out of play. This allows the superintendent to shift the underlying surface soil to a texture that can withstand the increased play.

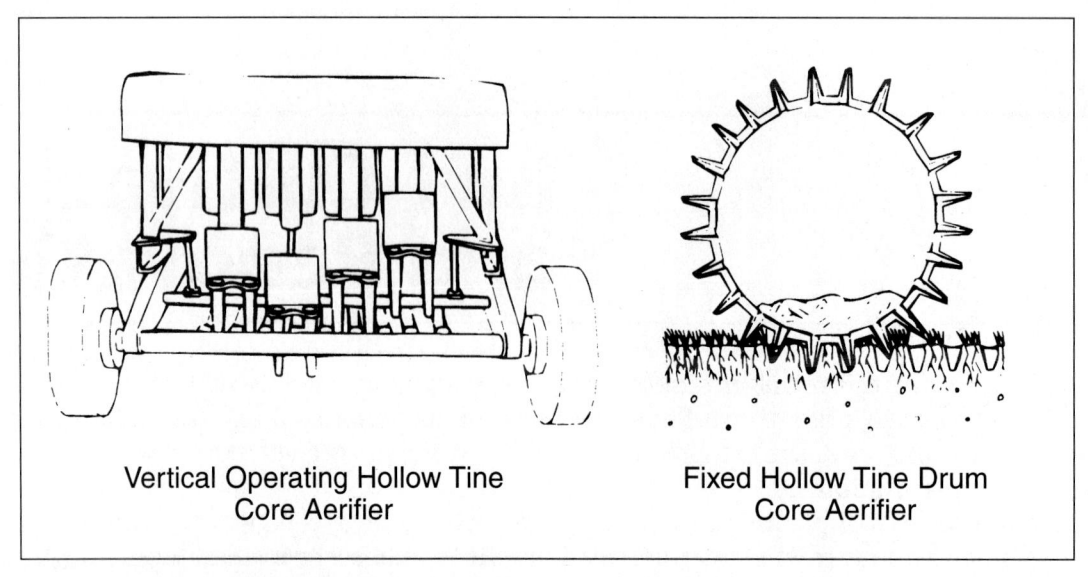

FIGURE 12-1 Two types of core aerifiers

Shatterhole Aeration

In some turf situations, there is a need to improve air and water movement into a turf, but, for various reasons, the turf manager cannot use standard core aeration. Shatterhole aeration involves driving a series of tines at a high force into the soil to create a shattering effect in the soil. This shattering or fracturing of the soil creates cracks for better water and air movement. The turf manager often uses a core aerifier with solid tines to create the shatter holes. Under the right soil conditions, many core aerifiers can provide the necessary force to create a shattering effect.

A major problem with shatterhole aeration is the potential for increasing compaction at the bottom and sides of the hole. And because there is no soil core removal, there is no relief from any compactive forces. Shatterhole aeration of moist loamy or clayey soils (which are already high in water content) can increase compaction problems.

To decrease the potential for compaction problems, new aerifiers are now available that use high speed "bullets" of water in place of solid tines. As these water bullets enter the soil, there is an "exploding" effect that churns the soil just beneath the turf as the water dissipates through the soil. These machines are very useful on golf course putting greens and sports fields because the machines create very small entry holes that do not disrupt the playing surface.

Spiking and Slicing

Spiking and slicing are very similar cultivation practices. Spikers have small short spikes, while slicers consist of thin triangular blades or thin discs. Both machines create shallow holes through a turf into the surface of the soil. Of the two, the turf manager will likely use slicers more than spikers.

Slicers are sometimes used to sever the lateral stems of creeping-type grasses, particularly creeping bentgrass and hybrid bermudagrass, so as to increase shoot density and encourage a more upright growth habit. A more common use for slicers, however, is turf renovation.

A slicer-seeder consists of a slicer mounted with a seed hopper. As the slicer blade goes through a turf, seed dribbles into the resulting slits before the slits reclose. This permits placement of seed in the soil of an existing turf so as to achieve the seed-soil contact necessary for germination and growth.

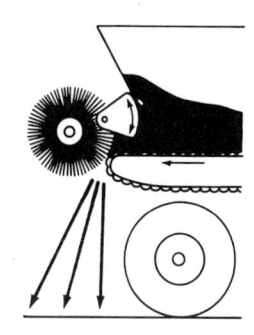

FIGURE 12-2 A mechanical topdresser

TOPDRESSING

Topdressing is the application of a fine layer of soil over an existing turf surface. Like some other cultivation practices, topdressing has many different uses. The turf manager often performs topdressing in conjunction with other cultivation practices, in particular, core aerification. Even though topdressing material can be applied by hand, a machine called a topdresser is often used to apply topdressing material (Figure 12-2).

Topdressing is most commonly performed on high quality sports turf. It is the major method for smoothing and leveling a playing surface, while at the same time helping to control thatch problems. On old golf greens, superintendents use sand topdressing in conjunction with core aerification to shift the surface soil to a coarser texture.

Because of the large amounts of soil and the associated handling and distribution required, topdressing is not a recommended method for most lawn situations. The only exception is for the leveling and smoothing of newly established turf areas. Most homeowners do not have the facilities or equipment to work with topdressing materials on an extensive basis.

THATCH CONTROL AND REMOVAL

Thatch is a layer of living and dead stems and roots between the green vegetation and the soil surface (Figure 12-3). Thatch is commonly associated with the creeping-type grasses, such as Kentucky bluegrass, creeping bentgrass, fine fescue, hybrid bermudagrass, and zoysiagrass.

Problems, Causes, and Benefits of Thatch

The problems of a very thick thatch layer include:

- poor rooting
- **scalping** from mowers
- localized dry spots
- an environment favorable to insects and disease organisms
- restricted fertilizer and chemical movement into soil

FIGURE 12-3 Cross-section of a turf showing thatch layer

Before ripping out all the thatch in a turf, one must be aware that there are some benefits to maintaining a moderate thatch layer. These benefits include:

- more resiliency on sports fields
- improved wear tolerance
- insulation against temperature extremes
- barrier against weeds

There are several causes of thatch, including:

- vigorous growing cultivars
- acidic soils
- poor soil aeration
- excessive water-soluble nitrogen levels
- infrequent mowing or excessively high mowing

Many of these causes can be controlled by adjusting cultural practices.

Because there are benefits to a moderate thatch and some thatch control methods can create additional stresses for the turf, the turf manager should not perform thatch control as a routine practice. In order to determine whether thatch is a problem the turf manager cuts a wedge of turf and examines the thatch layer. For lawns, if the layer is greater than one-half inch (1.27 cm) thick, the turf manager or homeowner should perform one or more methods of thatch control.

Biological Control of Thatch

Biological means of thatch control are efficient and result in the least amount of disruption and stress on a turf. The key is to maintain growth of turfgrass so that soil organisms can decompose the dead component of a thatch layer faster than the plant can add new tissue. Controlled fertilizer applications and other cultural practices to maintain growth help keep the thatch layer in check.

Because soil organisms control the dead component of a thatch layer, the turf manager must maintain soil conditions favorable to the organisms. Frequent, light liming helps to maintain pH near the neutral range, which is optimal for most soil decomposers. Aeration is another key factor in maintaining a stable population of soil decomposers.

Controlled irrigation combined with adequate drainage helps maintain an aerobic environment. An additional means of improving conditions is through core aeration. Although the practice appears to be more of a mechanical method and fairly disruptive to the turf, core aeration impacts less than 5 percent of the total turf surface. Core aeration creates favorable soil conditions for decomposing organisms while at the same time provides new avenues for turfgrass root growth. Topdressing also helps to create an environment favorable to decomposition. In addition, topdressing results in a mixing of soil with thatch to create a **mat** that not only creates a favorable environment but helps to inoculate the entire thatch layer with soil microorganisms. For many turfs, the topdressing material is leftover cores from core aeration. The turf manager uses a drag mat or **vertical mower** to break up the cores so the soil can sift down into the thatch layer.

Mechanical Control of Thatch

Thatch can be controlled with vertical mowers, power rakes, or dethatchers. These machines cut, rip, or tear thatch out of the turf. Because of the tearing action involved, the turf must have good rooting before any mechanical thatch removal is performed. Too often, an anxious homeowner uses a power rake to dethatch a lawn only to find when removing the debris that, in the process, the power rake pulled up all the turfgrass.

Because of the stress on the turf, mechanical thatch removal should be done only on an "as-need" basis. Furthermore, mechanical thatch removal should be performed when the turf is actively growing and can recover from the stress created by the practice. To avoid potential weed problems, the turf manager should refrain from removing all the thatch and exposing the soil surface. If the thatch is fairly thick, several shallow passes may be required to reach the desired layer thickness.

RENOVATION

Over time, a turf, for various reasons, will thin or shift in grass species. If the underlying soil is in good condition, the turf manager may opt to renovate rather than completely **re-establish** the turf. Renovation involves minor surface repair and overseeding into an existing turf stand. The existing turf stand may be left alive or killed with a non-selective herbicide but left in place.

Seeds must be in contact with the soil to germinate. Various cultivation machines can be used to open a turf for proper seed placement. The machines most commonly used are slicers, core aerifiers, and vertical mowers. For winter overseeding of bermudagrass, many southern turf managers use a groover. The groover is similar to a vertical mower but with thick blades and a heavy duty engine for cutting through the tough bermudagrass thatch.

Because renovation is a type of establishment, the turf manager must meet all the important factors for germination, such as environmental conditions, in order to achieve success.

ACTIVITIES

Turf Core Aeration

Equipment.
- core aerifier
- measuring tape
- calculator

Procedure.
1. Measure out a small turf area to be aerified.
2. Aerify the area with one pass of a core aerifier.
3. Calculate the total top surface area of the holes created by the core aerifier. (For this you will need to know the surface area of core holes and the number of core holes in the sample area.)
4. Calculate the percentage of total turf area consisting of core hole area. (For this you will need to know the surface area of test site.)
5. Measure the depth of several core holes.
6. Calculate a percentage of average actual depth penetration to desired depth penetration.

Data:
Percent of surface area as core holes: _____%
Percent actual depth of penetration to desired penetration: _____%

Topdressing

Topdress an area of turf as assigned by your instructor. Calculate the amount of topdressing material needed to cover the assigned turf area.

After a few weeks, take core samples from the turf and observe where the topdressing material has settled.

Compaction Control and Reduction

Apply different cultivation methods to a compacted area such as a football field or campus foot-traffic path. After a few weeks, apply the physical soil tests of bulk density and water infiltration (see Unit 4) to determine the effectiveness of the different cultivation methods.

Mechanical Thatch Control

Use the following equipment to mechanically remove thatch from a turf:

- garden rake
- crescent rake
- rotary-mower-mounted dethatching tines
- vertical mower
- power rake
- flail-type dethatcher

Evaluate each method as to effective thatch removal and damage to turf. Take soil samples before and after the thatch removal activities to measure the effectiveness of thatch removal.

Renovation/Overseeding Equipment

Using different overseeding methods and equipment, overseed various turf areas on campus. After a few weeks, look for germinating grass seeds as an indication of the effectiveness of each practice.

- garden rake and hand seeding
- core aerator and drop spreader
- slit-seeder
- grooving-type overseeder

EXERCISES

Cultivation Records

For use in later exercises and activities, maintain a record of any cultivation practices performed on a turf area. Include such information as method of cultivation, equipment used, time required to perform the activity, materials used, and purpose of the cultivation practice.

UNIT 13

Turf Equipment

Turf equipment is the number one capital investment that the turf manager will make to maintain a turf. With the rising cost of labor, many turf managers have turned to newer, larger, more efficient equipment. Manufacturers are continually bringing out new equipment for the turf manager. The price tag on many of the new mowers may appear excessive until adjustment is made for the labor savings over the lifetime of the equipment. Often, when total cost is taken into account, the new, more expensive equipment may be far less expensive than the original equipment that was replaced by the new equipment.

To maintain a quality turf, equipment service and maintenance are very important. Reduced visual quality of a turf often can be attributed to dull rotary mower blades or unadjusted, dull reel mowers. The useful life of equipment is shorter with poor or inadequate service and maintenance. The original cost savings can quickly change to a large expense if the turf manager does not practice routine service as recommended by the manufacturer.

EQUIPMENT SERVICE AND MAINTENANCE

For every piece of turf equipment, from hand grass clippers to the largest multireel-powered mower, the turf manager should perform routine service and maintenance on the equipment to ensure peak operating efficiency. Most powered turf equipment includes a service manual from the manufacturer that details the routine service and maintenance requirements of the equipment.

The turf manager should make an effort to maintain service manuals in a centralized location to avoid possible loss. If the service manual is missing from used equipment, the turf manager should make every effort to obtain a copy of the manual from the manufacturer or other source.

Mower Maintenance

As mowing is the major routine practice of any turf facility, sharp, well-adjusted mowers are important in maintaining a quality turf. Mowing not only has an impact on visual quality but also affects the playing quality of sports turfs and has an impact on the general health of any turf.

In most instances, reel mowers require more maintenance than rotary mowers. Because the reel mower cuts by a "scissoring" action, the need for precision cutting edges on the reel blades and the bedknife is much more critical than with a rotary mower blade.

Reel Mower Sharpening. Because of the numerous edges on the blades, reel mower sharpening is a time consuming task that requires several pieces of equipment to achieve a series of uniformly sharp precision cutting edges. The mechanic or turf manager must dismantle the mower to sharpen the reel blades on a reel grinder and sharpen the bedknife on a separate bedknife grinder. The mechanic will then reassemble and backlap the mower to obtain a matched, finished edge. Newer automated grinders reduce the amount of time a mechanic has to give direct attention to the sharpening process.

Backlapping involves attaching a mower to the reel so that it spins backwards. While the blades spin, a grinding compound is applied to the reel blades to hone the blades and bedknife to a matching fit.

Because of the work required for sharpening, the turf manager will usually have the mowers completely sharpened only once a year. Greensmowers are the exception, with golf course superintendents often sharpening them two to three times a year. However, the turf manager will do backlapping on a routine basis to maintain a sharp edge on the reels.

For some turf managers, the cost of buying, storing, and maintaining reel mower sharpening equipment far exceeds the cost of using the equipment. It often is more cost effective to have complete sharpening done by the service department of a turf equipment dealer. The turf manager then only does occasional backlapping to maintain the sharp cutting edges.

Rotary Mower Sharpening. Sharpening rotary mowers is a much easier task for the turf manager than sharpening reel mowers. One of the chief reasons for the popularity of the rotary mower is the ease of sharpening the blades.

The only pieces of equipment needed to sharpen rotary mower blades are a bench grinder with a medium grit or fine grit grinding stone and a balance point. Acceptable tools for most homeowners include a nail driven partially into a wall and a metal file or grinding stone attached to a drill.

Sharpening a rotary mower blade involves grinding the two cutting edges on the blade. In the process of grinding, blade balance must be maintained. If the blade becomes unbalanced, metal must be ground from the blade until it becomes balanced again. The purpose of balancing a rotary mower blade is to reduce vibration of the mower both for the sake of the operator and to prevent damage to any bearing assemblies.

Many turf managers who have several rotary mowers in constant use keep a large supply of replacement blades in stock. This way the mechanic or mower operator can change blades on a routine basis. Dull blades are stockpiled until a later time when the mechanic can sharpen all the blades at one time.

Fertilizing Equipment

Fertilizer equipment is fairly easy to maintain and service. The most important servicing to be performed on fertilizer-application equipment is keeping the equipment clean. Granular spreaders should be thoroughly rinsed and dried to prevent corrosion from fertilizer and pesticide materials. Sprayers require thorough rinsing and draining to prevent corrosion and to prevent blockage of the hoses and nozzles. Any engines and pumps need servicing on a regular basis, as recommended by the manufacturer.

Clean, well-maintained application equipment is important in maintaining calibration settings. Corrosion of poorly cleaned equipment can change the calibration settings of the equipment.

Irrigation System

Irrigation systems often are neglected because of their somewhat invisible nature, being that they are underground. The turf manager often does not realize the importance of irrigation system maintenance until the system breaks down during a mid-summer drought.

For northern turf managers, winter shut down and spring startup are important maintenance practices that need to be performed yearly to ensure a long-lasting, dependable irrigation system. Southern turf managers must perform regular, routine maintenance to ensure that the irrigation system can operate almost continuously for nearly twelve months of the year.

Other Equipment

Numerous other types of equipment used in managing a turf need a variety of routine service and maintenance work. Service schedules will vary based on the use of the equipment in a turf facility. On golf courses and large collegiate grounds, trucksters are probably the most often-used equipment and, thus, require constant service.

For equipment used only once or twice a year, it is important that the turf manager provide proper storage service on the equipment so it will operate when needed.

EQUIPMENT RECORD KEEPING

Maintaining records of all service and maintenance is an important part of equipment operation. A well-maintained equipment service record allows the turf manager to more efficiently schedule routine maintenance of the equipment. Maintenance records give indications as to the cost of maintaining equipment and help in deciding when to replace older or inefficient equipment.

Along with routine service records and repair records, additional records include fuel and lubricant usage logs. Fuel usage logs in combination with other information can provide the turf manager with an indication of general fuel costs. Also, if an underground fuel storage tank is on the site, any discrepancies between fuel pumped out of the tank and fuel usage could alert the turf manager to possible fuel loss through either a tank leak or theft.

REPLACING EQUIPMENT

Over time, even with proper maintenance and service, machinery will start to wear beyond repair. The equipment will require a greater amount of repair and service to keep it operational. The turf manager eventually must decide whether to replace the equipment by buying new equipment or to continue repairing the existing equipment. Repair of equipment is an operational expense that will increase as the equipment gets older and more used. There are several ways of deciding whether and when to repair or replace.

Fixed-Schedule Replacement

One method of avoiding increasing repair costs is by replacing the equipment on a fixed schedule or period. After a set number of years, the equipment is replaced with newer equipment. The old equipment is reassigned to other uses or sold as used equipment.

How long a period to wait before replacing equipment will depend on the equipment, its use in managing a turf area, and financial factors. Mowers used on golf courses or by commercial lawn maintenance companies need to be replaced much faster than mowers used in a typical home lawn or corporate lawn situation. Because of the infrequent use of some types of cultivation equipment, the turf manager may never need to replace this equipment. Also, manufacturers design equipment for many years of use. An irrigation system, if properly designed and maintained, can be used twenty years before there is a need to replace a worn pumping station.

A potential disadvantage of fixed-schedule replacement is when the turf manager needs to replace the equipment before the planned replacement date. Poor quality, abuse, or misuse of the equipment can all lead to a shorter life expectancy long before the budgeted monies become available for buying new equipment.

When replacing equipment, there is the matter of what to do with the old equipment. For some turf operations, particularly golf courses and large grounds facilities, turf managers will often relegate the older equipment to other uses. A good example is a golf course superintendent converting older triplex greensmowers into tee mowers when newer triplex greensmowers are purchased. In other situations, the superintendent will sell the old equipment and put the proceeds toward purchasing new equipment.

Equipment leasing is a form of fixed-schedule replacement. Leases are rentals of equipment over a fixed time ranging from one to five years. At the end of the lease period, the turf manager has the option of returning the equipment to the dealer or buying the equipment for the residual price. When returning equipment at the end of the lease period, the turf manager usually leases a new piece of equipment for another fixed time period.

There are several advantages and disadvantages to leasing turf equipment. A turf manager deciding whether to buy or lease should work with an accountant to determine which method is most economically feasible for the business.

Repair-Cost Breakpoint

Another way of determining whether and when to replace equipment is to use a repair-cost breakpoint. New equipment is purchased when the anticipated cost of repairs exceeds a breakpoint factor related to the replacement cost. An example of a cost breakpoint is when the anticipated repair cost exceeds 50 percent of the new equipment price. Another breakpoint might be the depreciated value of the equipment combined with anticipated repair costs.

The availability of capital funds at the time of anticipated replacement will have an impact on whether the turf manager should continue making repairs or buy new equipment. As with fixed-schedule replacement and leasing, the turf manager should seek the assistance of a financial analyst or accountant to determine which method works best in determining a repair or replacement cost breakpoint.

ACTIVITIES

Sharpening and Adjusting Mower Cutting Units

Observe and, if possible, participate in the techniques for sharpening both rotary and reel mower blades.

Comparison of Sharp Versus Dull Rotary Mowers

Equipment.

- rotary mower with a sharp blade
- rotary mower with a dull blade
- turf area

Procedure.

1. Divide a turf area into two equal areas. Use a different mower to mow each section, and compare the quality of cut. The mowers should be as follows:
   - Area 1: rotary mower with sharp blade
   - Area 2: rotary mower with dull blade

2. Apply visual quality ratings to the mowed areas, comparing the effectiveness of mowing by the different mowers. Enter the ratings in the comparison of mower maintenance on turf quality table (Figure 13-1).

	Area 1	Area 2	Area 3
Grass Species in Stand			
Rotary with Dull Blade			
Rotary with Sharp Blade			
Mis-adjusted Reel Mower			
Adjusted Reel Mower			

FIGURE 13-1 Comparison of mower maintenance on turf quality table

Comparison of Adjusted Versus Misadjusted Reel Mowers

Equipment.

- properly adjusted reel mower
- reel mower with misadjusted bedknife (gap increased between the bedknife and blades)
- turf area

Procedure.

1. Divide a turf area into two equal areas. Use a different mower to mow each section, and compare the quality of cut. The mower should be as follows:
   - Area 1: adjusted reel mower
   - Area 2: misadjusted reel mower

2. Apply visual quality ratings to the mowed areas, comparing the effectiveness of mowing between adjusted and misadjusted reels. Enter the ratings in the comparison of mower maintenance on turf quality table (Figure 13-1).

Equipment Record Keeping Systems

From vendor demonstrations, demonstration software, or actual software applications, evaluate commercially available equipment-inventory and record-keeping systems.

EXERCISES

Turf Mower Equipment File with Service Requirements

Modify your mower equipment file from Unit 7 to include service and adjustment requirements as they pertain to maintaining a sharp, well-adjusted mower.

Maintenance Scheduling and Record Keeping

Develop an equipment record-keeping system for tracking maintenance performed on turf equipment. Set up a calendar as a reminder for routine service and maintenance of equipment.

Using Paper. Develop a paper equipment record-keeping system using sample forms as suggested in a turf textbook, equipment maintenance textbook, or trade journal article. Use a large wall calendar to post notes reminding you when to perform service on a piece of turf equipment.

Using a Computer. Develop the equipment record-keeping system using a database manager. If the database manager or similar software supports "alarms," use alarm software to set up reminders notifying the turf manager when to perform service on a piece of turf equipment.

UNIT 14

Turf Management Practices

A major part of the turf management course involves the science and care of turfgrasses. In most turf situations, however, the practices of mowing, fertilizing, and irrigation are only part of the entire picture. It is important not only to determine how much fertilizer to apply and when to apply the fertilizer to the turf; but, also, to determine who is going to apply the fertilizer, how much the fertilizer is going to cost, and what impact (labor and financial) fertilizing will have on other turf care activities. **Turf management** is turf science and culture combined with the related business activities required to develop and keep a high-quality turf area.

Maintaining a quality turf area can be an expensive activity. To achieve a high-quality turf while keeping costs to a minimum, the turf manager should develop a management plan in addition to any cultural plans for the turf. The management plan serves as a foundation for developing a schedule of maintenance activities as well as a budget of proposed expenditures.

MAINTENANCE PLAN

Maintenance of many turf areas, particularly home lawns, often is a reactive practice. The homeowner mows the lawn when it needs cutting. There often is very little planning involved in mowing and other cultural practices. The turf manager, however, often oversees larger turf areas and a maintenance crew to care for the turf and landscape of those areas. Large sums of money and numerous pieces of equipment are used to maintain the turf. Furthermore, the turf manager often must answer to a higher authority responsible for the overall operation of a facility. The turf manager has a responsibility to develop a maintenance program as a way to increase operating efficiency and improve use of areas, facilities, and equipment while controlling costs and expenditures.

A management plan is not just a schedule of turf maintenance activities; it is a comprehensive document about the turf facility, how to maintain the facility. The basic contents of the management plan include:

- purpose of the turf facility
- area designation
- site inventory
- identified problems and solutions
- turf area objectives
- planned improvements
- standards and activities
- required equipment, material, and staff

Purpose of the Turf Facility

The turf manager, working in conjunction with superiors, should establish a statement describing the purpose of the turf facility. In some situations, there is a need to justify a particular turf facility. To some people, turf is a wasteful energy-and-money-sink that would be better used as a stand of trees or new building site.

The purposes will be as variable as the turf site. For a golf course turf, the purpose is to provide a place to play the game of golf. The purpose of a roadside utility turf, on the other hand, is to provide a safety zone along roads and highways while controlling dust and other particulate matter.

Area Designation

The area designation defines the boundaries of the area for which the turf manager is responsible.

Site Inventory

The site inventory defines the dimensions of all the turf areas as well as all buildings, hardscapes, and landscape features. The site inventory is important in developing maintenance activity plans, schedules, and budgets.

Identified Problems and Proposed Solutions

In developing a management plant, the turf manager should analyze the turf facility to identify both positive and negative site-maintenance factors. When problems are identified, the turf manager can explore various solutions.

The turf manager can use the turf facility analysis to set objectives that lead to the development of a comprehensive list of activities, schedules, and budgets.

Turf Area Objectives

Objectives are general statements of actions that the turf manager will perform to meet the goals and purpose of the turf facility. Objectives should have a recognizable outcome.

List of Planned Improvements

The turf facility analysis, may prove the turf facility to be in fairly good condition. The turf manager should, however, look for ways in which the turf facility can be improved. In some situations, improvements are not to correct existing problems but, instead, to prevent anticipated future problems.

Standards and Activities

Standards are extents or measurable factors to achieve the goals and objectives of a turf. An example of a standard would be "all golf greens will be maintained to provide a stimpmeter reading of 12 feet." For a grounds manager, a standard would be "all campus lawns will be mowed between 2.5 inches and 3 inches of cutting height."

Required Equipment, Material, and Staff

The final portion of the plan lists all the equipment, material, and employees needed to achieve the goals and objectives of the turf facility. This information comes from several different sources including the general cultural requirements of a turf and records of past turf management activities.

The management plan will vary based on the size and complexity of the turf facility. Roadside utility turfs will require a very simple management plan—one that will be adequate to keep the taxpayers happy. A management plan for a golf course will be fairly comprehensive because of the diverse turf areas. In addition, the turf manager is subject to additional demands because of the nature of golf and, particularly, the interest of golfers in having a quality playing facility.

TURF RECORD KEEPING

The foundation of a well-developed management plan and budget is a comprehensive set of records. These records detail the history of activities on the turf facility. Using these records, the turf manager can develop an accurate maintenance plan and budget.

Records are necessary for the evaluation of various turf maintenance activities. From well-kept records, the turf manager can determine the effectiveness and cost efficiency of various activities. The turf manager can then provide justification to a superior regarding the continuation or elimination of certain activities.

In addition to records being a foundation for plans and budgets, many regulatory agencies now require the turf manager to keep certain records. Employment records, accident and safety records, and chemical-activity records are the three major types of records that the turf manager must keep for several years.

The type of records that the turf manager keeps will vary depending on several factors including:

- size of the facility
- number of employees
- requirements of the organization
- accounting practices
- federal, state, and local regulations

The number and type of records will vary greatly between different organizations. Care must be taken to avoid excessive record keeping, which can tax the limited resources of the turf manager's staff.

TURF MAINTENANCE SCHEDULE

A turf maintenance schedule is a tentative plan of activities over time. The schedule can be subdivided into annual, monthly, weekly, and daily schedules. These schedules detail the activities to be performed within the given time period.

Many different factors impact the turf maintenance schedule, including the growing season, turf cultural practices, the use of the turf, the turfgrass, budget, personnel, and equipment. A detailed daily schedule of mowing and related activities will be applied to many golf courses. In contrast, a campus grounds may have mowing scheduled on a weekly basis.

When developing a turf maintenance schedule, the turf manager needs to build in some flexibility to allow for change should the need arise (such as in the case of severe weather conditions or equipment breakdown).

TURF BUDGET

A budget is an estimate of the cost to maintain a turf area. A budget enables the turf manager to control cost and to plan the operation of the turf area. The turf manager uses a budget to identify and control:

- all expenses
- the means and methods of cost reduction or containment
- the human factors in order to maintain

Most turf facility budgets contain two major categories: operating budget and capital budget. An operating budget covers the expenses to manage a turf for a predetermined time, usually the upcoming year. A capital budget covers expenditures for items and activities that will last longer than the length of the normal operating budget, which is usually one year.

Budget preparation involves:

1. Review of plans, goals, and objectives
2. Evaluation of present techniques and practices
3. Review and evaluation of records of past activities
4. Determination of needs for upcoming budget cycle (year)
5. Adjustment for various factors such as inflation and other activities

A well-developed management plan and schedule combined with efficient record keeping will allow the turf manager to quickly develop an accurate budget for the upcoming budget cycle.

Operating Budget

An operating budget is an estimated cost to maintain a turf during one budget cycle, usually one year. The turf manager will often split the operating budget into several different budget categories, the totals of which combine to make the complete budget estimate. These categories can be classified into three basic groups, *labor expense* (salaries and wages), *materials costs,* and *service expenses* necessary to maintain the turf. The budget for labor expense often is developed separately to take into account expenses beyond the wages or salaries paid directly to employees. Often, the turf manager will divide the materials costs and service expenses into more detailed categories such as fertilizers, chemicals, seed, fuel, equipment service, and utilities.

Capital Budget

The capital budget is an estimate of the cost of items and activities that will extend beyond the normal budget cycle, which is usually one year. Equipment such as mowers and the irrigation system are part of the capital budget. The turf manager sometimes will place major construction projects in the capital budget (although some turf facilities place construction projects in a separate budget).

TURF MANAGEMENT SYSTEMS

Many turf managers are turning to turf management systems to aid them with increasing record-keeping requirements and time-consuming analysis of the records. These systems are a combination of computer hardware and software for the purpose of record keeping, analysis, budgeting, and forecasting of a turf facility.

The systems range from simple accounting programs adapted for use by turf managers to comprehensive applications to handle all the information management tasks of the turf manager. Advanced systems can aid the turf manager in performing various cultural practices. They also can assist with required management activities through the integration of time clocks and on-site weather stations to track employees and monitor for pest problems.

ACTIVITIES

Observation of Actual Turf Management Systems

Visit with or attend presentations by various turf managers on the techniques they use in developing a management plan, schedule, and budget.

Turf Management Systems

With the assistance of your instructor, evaluate through demonstration software, demonstration by a vendor, or by use of actual applications, the various turf management software systems currently available to the turf manager.

EXERCISES

Campus Record System

Many of the activities and exercises in the past units included keeping records of turf practices on campus. Organize these records into a comprehensive record keeping system. Continue to keep records of the various activities performed on campus turf areas including costs. Use the information gathered in the record files as aids in developing a management plan and budget.

Using Paper. Reorganize your three-ring binder into a record keeping system that you can use later in your career. Include any additional forms with instructions.

Using a Computer. Use a combination of a word processor, spreadsheet, database manager, and other software tools to develop a comprehensive set of electronic documents.

Campus Turf Management Plan, Schedule, and Budget

As individuals or instructor assigned groups, develop an annual management plan, schedule, and budget for a designated area of campus, the entire campus, or a similar turf facility. Use the information developed and collected in previous activities and exercises as resources in developing the documents.

Using Paper. Develop a comprehensive set of documents. All documents should follow standard business practices that are typed and properly formatted. Integrate information and reports collected from previous activities and exercises.

Using a Computer. Use a combination of a word processor, spreadsheet, database manager, and other software tools to process information gathered in previous activities and develop a comprehensive set of electronic documents. If working in instructor assigned groups and networked workgroup computer resources are available, take advantage of workgroup software that allows a group of students to work on a single comprehensive set of electronic documents.

UNIT 15

Turf Communications and Publications

The prime purpose of the turf management course is to learn about the science, care, and management of turf areas. Knowledge of turf is often of little or no use, however, if the turf manager cannot communicate this knowledge to employees, customers, peers, and superiors. Agronomists with the U.S. Golf Association Green Section list communication, more specifically lack of proper communication, as the number one way the turf manager can fail in golf course management.

Communication is the transfer of information between two individuals or among many individuals. Methods of communication are speaking, gesturing, listening, writing, drawing, and reading. One can group these methods of communication into oral communication, that is, speaking, gesturing, and listening, and written communication, that is, writing, drawing, and reading. Turf management involves the use of all of these communication methods almost on a daily basis. Oral communication is the most prevalent with the turf manager. Turf management involves working with people: giving directions to employees, listening to employees' and clients' concerns, and, making presentations to superiors, clients, and peers. Superiors, clients, vendors, and peers also expect turf managers to provide written communications consisting of business letters, memos, and reports, as well as articles, brochures, and newsletters.

ORAL COMMUNICATION

Oral communication is the most common method of communicating with employees and customers. The turf manager is in fairly constant contact with employees on a daily basis. Most communication with employees is through oral directions and interactions.

Daily Communications

Even though the turf manager can give many instructions to employees in written form, oral directions allow for immediate communication to the employee. Oral interaction with employees allows for clarification of any instructions.

Presentations and Speeches

The public considers the turf manager to be a specialist who has a unique knowledge of turf. Groups ranging from the local garden club to trade associations often call on turf managers to give presentations or speeches on topics of turf care and management.

Even when turf managers do not give educational presentations to groups, they will still need to give presentations as part of the job. In many situations, the turf manager must give presentations to a supervisory board such as the green committee of a golf course or to a city parks board. The presentation could be an annual review of activities or justification of new budget expectations.

The turf manager must be able to speak in front of an audience. As with any other task involved in caring for a turf, giving presentations requires preparation and practice. The "dreaded speech course" that many turf curriculums require for graduation is one means by which a student as a future turf manager can learn the techniques of giving speeches and presentations.

WRITTEN COMMUNICATION

Although most daily communication, particularly with employees, is oral in nature, there are numerous times each day when the turf manager must communicate in writing. Letters, memorandums, proposals, and reports are typical daily written communications.

In addition, editors of local newspapers, garden club newspapers, or association trade journals may ask the turf manager to write an article for their respective publications. In many turf management situations, a newsletter is an effective means of promoting the activities of the turf manager and the turf maintenance crew. Brochures and similar promotional materials, including a customer newsletter, are important to the turf manager involved in a commercial service firm such as lawn care applicator or lawn maintenance. These tools are important in attracting and keeping customers.

ACTIVITIES

Presentations

Give a presentation to the entire class on a topic of turf management as assigned by your instructor or personally selected. When preparing the presentation, focus on preparing for a target audience that you will meet on the job. The target audience could be homeowners, employees, or other turf managers.

The presentation should last approximately 15-30 minutes, depending on the size of the class and other limits established by your instructor. The other members of the class or lab section should take on the role of the target audience, engaging you in dialog after the presentation.

Articles

Write one or more articles on a topic of turf management as assigned by your instructor or personally selected. Gear the articles toward a target audience that you will encounter on the job. The audience could be homeowners, employees, or peers. Articles could be for a variety of publications ranging from newspaper garden sections to industry trade magazines. Your instructor will act in the role of editor.

You should write the articles with a word processor so you can use the article files in later lab activities without the dreaded "retype." Also, when using a word processor, you can take advantage of writing tools such as spell checkers, grammar checkers, and readability evaluator software.

Brochures

Using various drawing packages, imaging systems (scanners), and desktop publishing (DTP) systems, compose a variety of promotional materials such as advertisements, notices, and flyers. Create materials that are commonly used by the lawn-care service industry to generate new customers. If possible, you should produce the flyers electronically for use in later lab activities.

Newsletters

With the assistance of your instructor, develop a newsletter combining the articles and compositions (brochure elements) from previous activities. You should gear the newsletter toward a target audience of people you will be serving in your career. This includes homeowners (for prospective lawn care specialists), club makers (for prospective golf course superintendents), your peers in trade associations, and community members (for public relations purposes).

Use a variety of mechanical composition tools, and computer tools, including word processors, drawing packages, and desktop publishing software, to generate the newsletter.

EXERCISES

Newsletter Articles

Write an article on a turf topic for submission to a garden club newsletter. Given that the audience consists chiefly of home gardeners, the focus of the article should be a turf management topic of interest and benefit to this group.

Industry Presentation

If a state turf association or research foundation has a student forum, submit a proposal for presentation at the annual meeting.

APPENDIX A

Introduction to Computers

A computer in simple terms is an electronic machine capable of following a list of instructions to perform calculations at a very rapid rate. What is more important, a computer is a multipurpose tool that enables the turf manager to store and process information in ways not easily done with paper and pencil.

A computer consists of hardware, or the actual machine, and software, or a list of instructions or programs that control the computer. The combination of computer hardware and software along with some related components composes a computer system. A personal computer, or "PC," is small enough to fit on a desk. Many can not fit in the space of a notebook. For almost all turf-management-related activities, the PC will be the computer of choice.

COMPUTER SOFTWARE

The real key to a computer system is the software. Software is a general term for computer programs or computer applications. These programs are lists of instructions that control the operation of a computer. Just as a cake recipe gives directions on baking a chocolate cake, a computer program gives directions to the computer on performing various calculations.

Without software, hardware is useless. It is the ability to control the machine through an infinite number of programs that gives computers their versatility. Software allows the turf manager to perform several types of information-management activities easily and quickly. The major types of software important to the turf manager are word processors, spreadsheets, and database managers. Telecommunications and graphics software are additional types of software that will be of interest to the turf manager.

Word Processors

The word processor is an important tool for the turf manager. As does any other business manager, a turf manager needs to write letters to customers and produce reports for upper management. Word processors are writing tools in the same category as pencils and typewriters. A word processor is very much like a typewriter. Instead of printing on paper, however, words appear on a computer display, or monitor, while being held and stored in the computer hardware. A major advantage of the word processor is the ability to easily correct mistakes that occur during typing.

Word processors have gone far beyond their original beginnings as "electronic typewriters." Word processors allow the turf manager to reuse previously written material without retyping the original material. For example, a lawn-care operator can personalize a "form letter" with very little additional typing. In addition, word processors now include extensive writing tools such as spell checkers and grammar evaluators to help eliminate common writing mistakes.

Today's word processors have features allowing production of documents that only expensive print shops could generate just a few years ago. This allows the turf manager to publish a variety of quality documents, such as newsletters, at substantially reduced costs.

Spreadsheets

The care of turfgrass requires a large number of calculations, to determine everything from fertilizer amounts to budgets. With the ever-increasing demand for precise applications of chemicals, it is important for the turf manager to perform calculations that provide accurate results. The spreadsheet, thus, is another important tool for the turf manager.

Spreadsheets contain a series of rows and columns where each intersection of a row and column is a cell. Numbers, text, or formulas can be placed in these cells. The formulas in a cell usually refer to values (numbers or text) in other cells. When a calculation is performed on a spreadsheet, formulas are processed and the result is displayed in the cell containing the formula. If a cell with a value changes, the spreadsheet automatically recalculates any formulas affected by the changed cell to give new results.

Spreadsheets allow the turf manager to answer "what if" questions such as, "What happens to my budget if the club manager tells me to cut 15 percent from all fertilizer and chemical purchases?" A spreadsheet-based budget containing fifty budget categories can recalculate in a matter of seconds to provide a new "bottom line" based on changes in a couple of categories.

Database Managers

Database managers provide the means to create an organized information management system. The foundation of many computer-based turf management systems and accounting systems is the database manager. It is the means to efficient record keeping for the turf manager.

In the simplest form, a database manager, or flat file manager, is the electronic equivalent of an index card file or rolodex file. A record in the database is the same as one index card, with the information contained in predefined fields. The turf manager can sort or arrange records in a database in a variety of ways based on the information in one or more fields. With searching methods, the turf manager can rapidly select a few records out of thousands.

Entering data into a database usually takes as much time as writing on index cards or in a notebook. The real strength of a database manager, however, is in the locating, summarizing, and analysis of information. A turf manager can quickly determine the sum of all chemicals applied in a year sorted by turf area. To do the equivalent analysis with paper records may take a few hours and be subject to errors.

Graphics Software

Graphics refers to pictures or drawings created on the computer. Graphics software is a diverse category usually subdivided into smaller categories. These categories include business graphics and charting, illustration, drawing, painting, and computer-aided design (CAD). Of the different categories, business graphics and CAD may be of most interest to the turf manager.

Business graphics, or charting, is very similar to the graphing done in math class. This software plots graphs to represent the relationship of numbers or groups of numbers. Charting programs are often part of spreadsheets or databases, which eliminates the need for re-entering data.

With a charting program, the turf manager can take a budget and convert it into a pie chart to show the relationship between the different budget categories. As an example, a turf manager can create a bar chart or line chart from the irrigation usage spreadsheet or database to quickly show increasing water usage during dry summer periods.

Although CAD often is used for landscape design the turf manager generally uses CAD as a mapping tool or visual-information manager. The drawing tools in a CAD program allow the turf manager to quickly create site plans for record-keeping purposes. From a CAD drawing, the turf manager can create a large variety of working paper drawings without damage to the original plans. CAD-based drawings are also easier to update as turf areas change.

Telecommunications

Telecommunications software enhances the turf manager's ability to communicate with other computer systems over various networks including the telephone system. The software includes such features as advanced file transfer, automated programming, and control of remote computer systems.

As an example, to find the latest article on controlling insect pests in turfgrass, a turf manager can have the software automatically call up an on-line information service later at night when phone rates are lower. Many computer-controlled irrigation systems allow the turf manager to operate the irrigation system via remote control using telecommunications software.

Software Programming

Although major software development should be left to professional software developers, there are many times the turf manager may wish to have the computer perform some additional tasks. Whether these tasks be repetitious data entry activities or turf-specific calculations, a turf manager can accomplish a number of tasks using the programmability features found in many major software packages.

These programmability features are either macros or application-specific programming languages. Although these features are not as sophisticated as the systems used to create software applications, they can be fairly powerful and involve the same programming methodologies found in advanced software development systems.

Fortunately, the turf manager does not need a computer science degree to start using these programming features. Macros are a group of commands that can be activated by typing a short keyword or "pressing" a button on the screen. Creating a macro involves activating a "macro recorder." As one issues various commands while using the software, the software saves each command for later use. Just like a video recorder, one can re-run recorded macros at any time.

More advanced software applications, including some word processors, database managers, spreadsheets, and CAD systems, contain advanced application-specific languages that allow the turf manager to alter the characteristics of the application. Many information management systems, including some for turf management, are standard database managers modified by the application-specific language contained in the DBMS to meet certain program objectives.

Software applications with application-specific languages are more sophisticated than macros, allowing the turf manager to quickly develop small applications or modify a software application to meet specific turf information management needs.

COMPUTER HARDWARE AND OPERATING SYSTEMS

Computer hardware is a combination of various components integrated together to allow for the input of commands, programs, and data; the processing of commands, programs, and data; the storage of data and programs; and the output of results. There are four major component groups: processor (central processing unit, or CPU), memory, storage, and input/output, or I/O.

Processor

The processor is the "brain" of the computer system. As the term infers, it is responsible for processing data by following instructions in a program. The CPU is another term for the processor. Many larger computer systems have additional processors to support the primary or central processor. Processors that in the past took up large rooms now consist of a single electronic component or "chip" no larger than a matchbook. This reduction in size has allowed manufacturers to provide a complete computer system to fit on a desk or in a notebook. Most personal computers contain only one processor or CPU.

There are various processors, each with different features and internal design. A software developer often will write programs or applications to take advantage of the *instruction set*, or features of a particular processor. Software written for one processor often cannot operate on a computer with a different processor.

Selecting software to meet a task and then selecting the processor required by the software will avoid later problems. Selecting a computer based on the processor is similar to buying a coal-fired steam tractor and finding coal is no longer available.

Memory

During processing, the processor needs space to work with data and programs, very much like a workbench or desktop provides the space to work on equipment or do paperwork. The term for this workspace is *memory*. The computer industry commonly defines the size of memory as bytes, usually in thousands of bytes, or kilobytes (KB), and millions of bytes, or megabytes (MB or "megs"). For ease of understanding memory size, one byte is equivalent to one character on a printed sheet of paper. A standard 8.5-by-11-inch, double-spaced, typewritten page holds about 2,000 characters or "2 KB of memory." A tabloid size (11-by-17-inch) sheet of paper takes about 4KB of memory.

Memory can further be divided into random-access memory (RAM) and read-only memory (ROM). The CPU uses RAM as the workspace for temporary holding of data and programs. Random access memory is volatile and capable of being changed. Loss of electrical power will result in loss of anything in RAM memory.

Read-only memory, unlike RAM, does not lose data or programs when the computer loses electrical power. As soon as a manufacturer places anything into a ROM chip, it is there permanently. Changing the contents of a ROM chip requires replacing the old ROM integrated circuit chip with a new chip. ROM often contains data and software important in the general operation of the CPU and other components.

Storage

Because RAM is volatile and limited in size, there is a need for other means of storing data and programs that are not in use at any one time. There are several devices for storage, the most common being the magnetic disk drive. These drives store data on a magnetic media using methods similar to an audio or video cassette recording. The disk format, unlike linear tape, however, allows for quick access to data anywhere on the disk. There are two types of disk drives: floppy disk drives and hard disk drives.

A floppy disk contains a flexible magnetic material and works very much like an audio cassette, where the recording head contacts the media. A flexible jacket or hard shell encloses the floppy disk for protection. The major use of floppy disks is the transfer of software programs and data between computer systems. This includes the duplication of information from hard disk drives for archiving or backup purposes. Floppy disk drives found in most personal computers have a maximum capacity of 2 MG.

The main storage device on computers is the hard disk drive. A hard disk consists of one or more rigid platters coated with a magnetic material. The disk spins at a very high speed, with the recording heads "flying" just above the disk. This allows for very rapid data and program transfers between the disk drive and the CPU. It also allows for very large storage capacities. Some hard disk drives are now capable of storage sizes in the gigabytes (1,000 megabytes)—all in a space no larger than a turf management textbook.

Because of the rapid access associated with hard disk drives, portions of a hard disk drive can be set aside as virtual memory without incurring the expense of adding additional physical memory. With the right CPU and appropriate operating system software, a computer's logical memory can be increased by using the hard disk drive as a means of simulating memory.

The CD-ROM is another type of disk drive that is increasing in use. The term CD-ROM means "compact disc-read only memory." It uses the same technology as audio compact discs, where a laser beam reads encoded digital information. A CD-ROM can hold audio music tracks for play in a stereo system's CD player, and audio compact discs can be played in a CD-ROM drive. As the term infers, a CD-ROM is read-only. After the compact disc has been created, it cannot be erased. Like a phonograph record, however, a manufacturer can duplicate a CD-ROM at a lower cost than that required to duplicate magnetic media.

The CD-ROM combines the portability of a floppy disk with the capability of a hard disk drive (550 megabytes) to move large amounts of data or large program files. Because of the low duplication costs, the CD-ROM is rapidly replacing floppy disks for distribution of large data files. Because of its large capacity, a CD-ROM is able to store an entire twenty-six volume encyclopedia, including pictures, on one disk. Many reference services now provide their indexes on CD-ROM in place of multi-volume paper copies.

Input/Output

To interact with the computer, particularly the CPU, a means of entering data and programs is needed. As soon as the CPU finishes processing, a means of obtaining the results is needed. The most common input device is the keyboard which is very similar to an old typewriter keyboard except that it has additional keys. A mouse or similar pointer device (trackball, pen, touch screen, tablet, voice) is another common input device, which is used primarily for controlling the computer.

The two common means of output are video displays and printers. Video monitors use the same basic display technology as does a television, but without the channel selectors. Another display found mostly on notebook computers is the liquid crystal display or LCD. Most monitors display color, although some low-cost computer systems have monochrome monitors.

The printer provides a paper copy of the output. There are several different types of printers, ranging from dot-matrix to inkjet and laser. Each type has its advantages and disadvantages. Laser printers and inkjet printers are very popular due to the high quality of printing. Some of these printers are also capable of producing color output.

Network Interface

In order for a computer to communicate with other computers, a network interface is needed. The network interface connects a computer into a network of other computers, usually via the use of wiring or cable. With the appropriate software, the turf manager can then share files with other turf managers or control an irrigation computer from the home computer system.

Connecting computers within a building often involves the use of network interface connectors and cabling between the computers. This local area network (LAN) permits the turf manager to share files with other employees in the office or to control other computers within the turf facility, such as an irrigation control computer.

A wide area network (WAN) may be required to connect computers far away from each other. The telephone system is one type of WAN commonly used for networking computers. To connect a computer into a WAN such as the phone system requires a modem. This device converts computer data signals into a format that can be sent over common telephone lines.

Operating System

A final part of computer hardware is actually a special type of software. The operating system is the software that controls the basic operation of the computer hardware. The most basic operating system is the disk operating system, or DOS. This software is responsible for controlling disk drives. It has grown to include basic control of the input and output devices (keyboard, printers, monitors) as well as other hardware components.

More advanced operating systems control not only the basic operating of the hardware, but also programs running in the computer and the interaction with whoever is using the computer. These operating systems provide an extensive interface for operating the computer. Current technology for these advanced interfaces is graphic user interfaces, or GUIs; more appropriate terminology is common user access interface, or CUA.

The main feature of a CUA-based operating system is user interaction with the computer via a simulated desktop. On the video display are stackable, resizable "windows," which are like sheets of paper than can be shuffled around on a desk. The user can issue all commands to the computer through pull-down menus and pop-up dialog boxes. In addition, any program that takes advantage of this advanced interface has the same operating features (menus, windows, dialogs). Consequently, as soon as employees of the turf business learn one program, they have learned the basic operation of any program that runs under the CUA-based operating system. This is a major advantage to the turf manager in that it reduces training time by increasing the ease of use.

APPENDIX B

Using the On-Line Companion

The on-line companion that can be found at www.agriscience.delmar.com contains supplemental files to many of the exercises and some activities. These are basic "template" files to give you a start in completing the exercises. You are expected to enhance the files to complete the various exercises assigned by your instructor.

The author developed the files using Microsoft® Works 3.0 for Windows. To use the files without conversion requires you to have access to Works 3.0 for Windows or a later version. Unit 1 includes an activity for determining the software applications that will be available for later exercises.

For Microsoft Works for DOS and Microsoft Works for the Apple Macintosh®, as well as for other software applications, an additional set of files is available in an intermediate transfer format. Most of the general purpose software applications available on the market can open these intermediate-format files. Apple Macintosh users should read the following section on moving the files from a PC DOS disk format into a Macintosh disk format.

File Names

The file names are keyed to each unit. The name of the file related to each exercise or activity is listed like this:

On-Line Companion. U02E01.WDB—Turf Article Reference Database

The "U02" in the name refers to the unit; in the example, it refers to Unit 2: Turf Information Management. The "E01" refers to the exercise number, or Exercise 1. A few files have an "A" in place of the "E" to refer to a computer file used in an activity, such as "U05A01.WKS."

Some file names have an "A" or "B" following the exercise or activity number, such as "U02E03A.WKS" and "U02E03B.WDB." This is to represent files to the same exercise but in different application formats. In the example, you have the option of doing the exercise with a spreadsheet (WKS) or a database (WDB).

The three-letter extension after the period, such as WKS, WDB, or WMF, is a method used by MS Works to distinguish the type of file. It is a way for the application and you to determine which tool (spreadsheet, database, word processor) to use in MS Works for opening the file. The extensions represent the following types:

WKS	An MS-Works spreadsheet table
WDB	An MS-Works database file
WPS	An MS-Works word processor document
WCM	An MS-Works communications file
WMF	A Windows drawing file that MS Works can open

Using the Files with Microsoft Works for Windows

All the MS-Works-for-Windows files are in the root directory of the supplemental files. For instructions on working with the files in MS Works for Windows, consult the following resources:

The Microsoft Works User's Guide
The on-line Help inside the Microsoft Works application
The Works Tutorial inside the Microsoft Works application
The Cue Cards inside the Microsoft Works application
The Works Wizards inside the Microsoft Works application

See your instructor or computer lab support staff for additional assistance.

Using the Files with Other Applications and Computers

The subdirectory, OTHERAPS, contains the files in a translated format that you can open with many other applications. Consult the software application's documentation on opening files in a foreign format. Unit 1 includes activities for opening and accessing files in subdirectories. If you have difficulty, see your instructor or computer lab support staff for assistance.

Translated File Formats. The translated files are in the following file formats:

MS Works Extension	Translated File Extension	Notes
WKS	WK1	The Lotus 1-2-3 file format that can be opened by many other spreadsheet applications
WDB	DBF	The dBase III file format that can be opened by many other database manager applications
WPS	RTF	The Microsoft Rich Text Format for transferring files between different word processors
WCM	None	None
WMF	None	None

Using the Files on the Apple Macintosh

To use the files on the Apple Macintosh requires moving them from a PC DOS disk to a Macintosh disk. The three basic methods for accomplishing this are: Apple File Exchange, DOS mounters, and serial communications. The method used will depend on the Macintosh model and software. All Macintoshes produced after 1987 are capable of reading and writing to 3.5 inch, 1.44 megabyte, DOS-formatted diskettes.

Apple File Exchange. Apple File Exchange is a utility for moving files on DOS diskettes to Macintosh diskettes or hard disk drives. Information on using Apple File Exchange can be found in the user's guide included with the system software.

DOS Mounters. Several applications are available that will allow DOS diskettes to be used like Macintosh diskettes without having to transfer the files using Apple File Exchange. The latest system software supports this DOS-diskette-mounting feature. There are several commercial applications for Macintosh users with older system software. Consult a computer supplier for software availability.

Serial Communications. For older Macintoshes without the hardware or software to work directly with DOS diskettes, serial communications can be used to transfer files. Serial communications consist of cables or modems between a Macintosh and PC along with telecommunication software. It is best to seek the assistance of a computer wizard for serial transfer.

Supplemental Disk File List

The following is a list of Microsoft Works 3.0 for Windows files with brief descriptions.

U02E01.WDB	— Turf article reference database
U02E02.WMF	— Sample site drawing
U02E03A.WKS	— Site inventory spreadsheet
U02E03B.WDB	— Site inventory database
U03E01.WDB	— Turfgrass species database
U03E02.WDB	— Turfgrass cultivar file
U04E01A.WDB	— Site inventory with soil test results database
U04E01B.WKS	— Site inventory with soil test results spreadsheet
U05A01.WKS	— Seed mixture spreadsheet
U05E01.WDB	— Turfgrass species/cultivar selection database
U05E02.WKS	— Establishment rate spreadsheet
U06E01.WKS	— Fertilizer information spreadsheet
U06E02.WKS	— Fertilizer rates spreadsheet
U06E03.WKS	— Fertilizer cost analysis spreadsheet
U07A01.WKS	— Mowing time estimates activity spreadsheet
U07A02.WKS	— Mowing cost activity spreadsheet
U07E01.WDB	— Mower equipment database
U07E02.WKS	— Mowing time and cost analysis spreadsheet
U08A01.WKS	— Irrigation uniformity activity spreadsheet
U08E01A.WKS	— Soil moisture log spreadsheet
U08E01B.WDB	— Soil moisture log database
U08E02A.WKS	— Weather record spreadsheet
U08E02B.WDB	— Weather record database
U08E03A.WKS	— Water budget checkbook spreadsheet
U08E03B.WDB	— Water budget checkbook database
U09E01.WDB	— Pest problem database
U10E01.WDB	— Pesticide inventory database
U10E02.WKS	— Pesticide calculations spreadsheet
U10E03.WDB	— Pesticide application database
U11E01.WDB	— Materials calibration database

Glossary

A

Acidic soil. Soil with a pH below 7.0.
Active ingredient (ai). The actual toxic material present in a pesticide formulation.
Aeration or aerification. The practice of improving the soil by making holes or slits in it.
Alkaline soil. Soil with a pH higher than 7.0.
Auricles. Extensions of a grass leaf collar, which may or may not be present on all grasses.

B

Back lapping. A backward turning of the reel while applying a grinding compound to create matched surfaces between the reel blades and the reel bedknife.
Bedknife. The fixed bottom blade of a reel mower against which the spinning reel blades cut with a scissor action.
Bench setting. The height of the mower cutting blades above a solid surface such as a workbench or floor. Also known as *mowing height*.
Blade. The flat, upper portion of a grass leaf above the leaf sheath.
Blend. A combination of seeds comprising two or more cultivars of one turfgrass species.
Broadleaf. A plant with a broad, flat leaf and the net-like leaf-vein pattern common among dicot plants.
Bud. A cluster of meristematic (capable of division) cells that may develop into specialized plant tissues including stems, leaves, and flowers.
Bulk density. The density, in grams per cubic centimeter (g/cm^3), of an undisturbed soil as a quality indicator of soil structure.
Bunch-type Growth Habit. Plant growth by tillering, or the formation of non-lateral side shoots.

C

Calcium carbonate equivalent. The number of pounds of calcium carbonate needed to neutralize the acidity caused by a ton of fertilizer.
Cation. A positively charged ion such as potassium (K^+) and ammonium ($NH4^+$).
Cation exchange capacity (CEC). The quantity of exchangeable cations that a soil can absorb as an indicator of the soil's ability to store nutrients.
Chlorosis. A yellowing of green plant tissue due to lack of chlorophyll, which is often an indicator of nurient deficiency.
Collar. A narrow band defining the junction between a leaf blade and leaf sheath.
Compaction. An unfavorable soil condition where soil particles are pressed together, causing an increase in soil bulk density.
Complete fertilizer. A fertilizer with the three major mineral nutrients of nitrogen, phosphorus, and potassium.
Computer. An electronic machine capable of performing calculations at a high rate of speed under the guidance of a program or instruction set.
Cool-season turfgrass. Turfgrass species that grow best in the temperature range of 60°-75°F (15°-24°C).
Core cultivation. A method of cultivating the soil through removal of thin soil cores or plugs. Also known as core aeration of core aerification.

Coring. Another term for core cultivation; usually refers to the active practice.
Creeping-type growth habit. The ability to spread horizontally with rhizomes or stolons.
Crown. The major meristematic area, or growing point, of a grass plant, which is usually located near the soil surface.
Cultivar. A plant within a species that differs from other members of the species and retains these distinguishing features when reproduced.
Cutting height. The actual height above a soil surface at which grass is cut. Also known as the *effective cutting height* (as opposed to the *bench setting*).

D

Desiccation. The loss of moisture from a plant because of drying weather conditions or chemicals such as salt-type fertilizers.
Dethatch. To remove an excessive thatch layer with a machine such as a power rake or vertical mower.
Dicot, dicotyledon. A type of plant with two cotyledons, or seed leaves; often distinguishable by broad leaves with net-like leaf veins.
Disease. An abnormal condition usually caused by pathogens such as fungi, bacteria, viruses, or nematodes.
Dormancy. A resting stage when a plant's growth rate is greatly reduced or nonexistent because of extended cold, heat, or drought.
Dormant seeding. Planting seed after favorable growing conditions so that the seed remains dormant until favorable conditions recur to trigger germination.

E

Ecosystem, turfgrass. A turfgrass community, the surrounding environment, and all other organisms living in the surrounding environment.
Evapotranspiration (ET). The combined loss of water from the plant, as transpiration, and from the soil surface, as evaporation.

F

Fertigation. The application of liquid fertilizer through the irrigation system.
Fertilizer analysis. The percentage by weight of nitrogen, available phospohric acid, and soluble potash in a fertilizer.
Fertilizer ratio. The comparison of the amounts of nitrogen, phosphorus, and potassium in a fertilizer.
Formulation. The form of a chemical pesticide, such as emulsifiable concentrates, wettable powders, flowables, and granules.
Fungicide. A chemical used to control disease-causing fungi.
Fungus. A form of plant life that lacks chlorophyll and obtains its food from other organic sources, whether dead or living.

G

Granules. Solid particles of fertilizer or carriers for pesticides.
Grub. The larval or immature stage of various beetle insect species.

H

Herbicide. A chemical used to control undesirable plants in a turf.
Hybrid. The offspring of genetically dissimilar parents.

I

Infection. The invasion and establishment of a pathogen in a plant.
Inflorescence. The flower part of a plant where the seed develops.
Internode. The part of the stem between two nodes.
Irrigation. The application of water from sources other than natural precipitation to meet the water needs of the plant.

L

Larva. The immature version of insects that undergo complete metamorphosis.
Lateral growth. The sideward or horizontal growth of stems and shoots.
Leaching. The loss of materials, usually mineral nutrients, dissolved in soil solution by the downward movement of water through the soil.
Leaf texture. The width of the leaf blade.
Ligule. An extension of the epidermal layer located just in front of the collar at the junction of the leaf blade leaf sheath.
Lime. Calcium and magnesium compounds used to raise the soil pH (and, thus, decrease acidity).

M

Mat. A layer of intermixed thatch and soil.
Membranous. Having a thin, soft, transparent or skin-like tissue.
Metamorphosis. The transformation or change in appearance that occurs between the stages of an insect.
Micronutrient. Mineral nutrients required in very small amounts by the plant.
Mixture, seed. A combination of seeds comprising two or more species of turfgrasses; each species may be a blend.
Mole. A small subterranean tunneling animal that is very disruptive to turf.
Monocot, monocotyledon. A type of plant with one cotyledon or seed leaf; often distinguished by narrow leaves with parallel leaf veins.
Monostand. A turf area comprised of only one cultivar.

N

Nematode. Small, often microscopic, eel-like worms.
Node. The joint of a stem where the leaf is attached and buds occur.
Nonselective herbicide. A herbicide that kills many different types of plants.
Nymph. The immature version of insects that undergo incomplete metamorphosis; nymphs are similar in appearance to the adult version of the insect.

O

Organic matter. The material resulting from the growth of living organisms.
Overseed. The seed into an existing turf stand.

P

Pathogen. A disease-causing organism.
Percolation. The downward movement of water through the soil.
Perennial. A plant that has a life span of two or more years.
pH, soil. The measurement of the acidity or alkalinity of a soil.

Photosynthesis. The process by which green plants capture light energy with carbon dioxide and water to form carbohydrates, a form of stored chemical energy.
Plug. A small piece of sod used for vegetative establishment and turf repair.
Plugging. The process of establishing a turf with plugs.
Polystand. A turf area composed of two or more cultivars or species.
Post-emergent herbicide. A chemical used to control weeds after their emergence from the soil.
Pre-emergent herbicide. A chemical used to control weeds before their emergence from the soil.
Pure live seed (PLS). The percentage of seed that is pure and capable of germination.
Purity, seed. The percentage by weight of seed of each cultivar in a seed blend or mixture.

R

Recuperative ability. The ability of a plant to recover from injury.
Reel mower. A mower with a rotating reel of blades that cuts against a stationary bedknife.
Re-establishment. Replanting a turf area by destroying the old turf and tilling the soil to form a new planting bed.
Renovation. Replanting a turf area without tilling the soil; the existing turf may be left alive or killed with a nonselective herbicide.
Rhizome. A lateral-growing, underground stem.
Rotary mower. A mower that cuts by impact of a horizontal spinning blade.

S

Scalp. To mow a turf so close that all green leaf tissue is removed from the turf.
Selective herbicide. A chemical that kills only certain types of plants while causing little or no damage to other types of plants.
Sheath. The lower portion of a grass leaf that forms a tube-like structure around the stem.
Shoot. A stem with the attached leaves.
Sod. Strips of live grass with a small amount of adhering soil; used in vegetative turf establishment.
Soil. A dynamic body of inorganic and organic material on the earth's surface, capable of supporting life.
Soil texture. The relative proportions of various-sized soil particles.
Species. A group of organisms with common characteristics; capable of interbreeding to produce fertile offspring like the parents.
Spiking. A cultivation practice in which solid tines or spikes penetrate the soil for the purpose of improving air and water movement into the soil.
Spreader. A type of equipment used to uniformly distribute material over a turf area.
Sprig. A piece of stem, usually rhizome or stolon, used for vegetative propagation of certain grasses.
Sprigging. The practice of vegetatively establishing a turf by planting sprigs in furrows or holes.
Stem. The part of the plant with nodes and internodes that supports leaves and flowers.
Stolon. A lateral-growing, above-ground stem.
Stolonizing. The practice of vegetatively establishing a turf by broadcast planting sprigs over a turf area.
Systemic. A chemical absorbed into a plant through the leaves or roots; when inside the plant, the chemical moves through the vascular tissue to various parts of the plant.

T

Tensionmeter. A device for measuring soil moisture tension, used as a means of determining irrigation need.
Thatch. A layer of living and dead stem and root tissue between the zone of green vegetation and the soil surface.

Tiller. A stem that grows from the crown within the enclosing leaf sheaths.

Topdresing. To apply a fine layer of soil or similar material over an existing turf.

Transition zone. A zone where both cool-season grasses and warm-season grasses can grow.

Transpiration. The movement of water into the plant through the roots, upward in the plant, and back out of the plant as water vapor through the leaves.

Turf. A covering of mowed vegetation, usually grasses.

Turfgrass. Grasses used in a turf situation.

Turfgrass culture. The science and culture of turfgrasses used in a turf situation.

Turfgrass management. Turfgrass culture along with the related business practices necessary to maintain a turf.

Turfgrass science. The study of turfgrasses and how they grow in a turf situation.

V

Variety. Usually another term for *cultivar;* some taxonomists refer to a variety as a plant with naturally occurring characteristics as opposed to a cultivar, which has man-induced characteristics.

Vernation. The arrangement of leaves, either rolled or folded, as they grow up through the leaf sheaths of older leaves.

Vertical mowing. Slicing into a turf with vertically mounted, sharp knives for the purpose of thatch control.

W

Warm-season turfgrass. Turfgrass species that grow best in the temperature range of 80°-95°F (27°-35°C).

Reference Library

Ali, A. D., and C. L. Elmore. *Turfgrass Pests.* Cooperative Extension, University of California Division of Agriculture and Natural Resources, 1989.

Beard, J. B. *Turfgrass Management for Golf Courses.* Macmillan Publishing, 1982.

Beard, J. B. *Turfgrass: Science and Culture.* Prentice-Hall, Inc., 1972.

Bormann, F. H., D. Balmori, and G. T. Geballe. *Redesigning the American Lawn: A Search for Environmental Harmony.* Yale University Press, 1993.

Daniel, W. H., and R. P. Freeborg. *Turf Manager's Handbook.* Harcourt Brace Jovanovich Publications, 1979.

Danneberger, T. K. *Turfgrass Ecology & Management.* G.I.E. Inc., 1993.

Decker, H. F., and J. M. Decker. *Lawn Care: A Handbook for Professionals.* Reston Publishing Co., 1988.

Emmons, R. D. *Turfgrass Science and Management, Third Edition.* Delmar, 2000.

Hanson, A. A., and F. V. Juska (eds.). "Turfgrass Science," *Agronomy Monograph No. 14,* American Society of Agronomy, 1969.

Madison, J. H. *Practical Turfgrass Management.* Van Nostrand Reinhold Co., 1971.

Madison, J. H. *Principles of Turfgrass Culture.* Van Nostrand Reinhold Co., 1971.

Shurtleff, M. C., T. W. Fermanian, and R. Randell. *Controlling Turfgrass Pests.* Reston Publishing Co., 1987.

Turgeon, A. J., and F. Giles. *Turfgrass Management.* Reston Publishing Co., 1980.

Waddington, D. V., R. N. Carrow, and R. C. Shearman (eds.). "Turfgrass." *Agronomy Monograph No. 32,* American Society of Agronomy, 1992.

NOTES